DATE DUE

1992

THE JOHNS HOPKINS OCEANOGRAPHIC STUDIES

Number 2

The Johns Hopkins Oceanographic Studies, published by The Johns Hopkins Press, is a series designed for the publication of monographs and long technical papers in the field of oceanography. Manuscripts may be submitted to any member of the Editorial Board, which has the responsibility for appraisal and selection of papers for publication.

INTERMEDIATE WATERS OF THE PACIFIC OCEAN

THE JOHNS HOPKINS OCEANOGRAPHIC STUDIES

Intermediate Waters
of the Pacific Ocean

by

Joseph L. Reid, Jr.

Scripps Institution of Oceanography
University of California
La Jolla, California

THE JOHNS HOPKINS PRESS: BALTIMORE

1965

Contribution from Scripps Institution of Oceanography,
University of California, San Diego

ABSTRACT

In the Pacific Ocean two great tongues of low-salinity water extend equatorward from the surface in high latitudes to the intermediate depths in middle and low latitudes. The tongue from the north is lower in salinity and lies at shallower depth than that from the south, and they overlap in a small area north of the equator.

Because of these tongues the intermediate-depth waters of low and middle latitudes are lower in salinity than the waters above and below. The sources of this low-salinity Intermediate Water are apparently the high-latitude areas of precipitation, but the means of passing this characteristic property from high latitudes and dissipating it in low latitudes has been obscure.

Earlier studies of the Intermediate Water of the North Pacific had assumed or concluded that it is formed at high latitudes in the mixed layer in winter and sinks to greater depths as it flows toward lower latitudes, either as an undercurrent along the western boundary or as a part of the general wind-driven circulation of the subtropical anticyclones. Although the subsequent effects of the horizontal circulation, the vertical mixing and transport, and the lateral mixing had been studied, the original assumption—that the Intermediate Water was formed in the mixed layer—had been retained.

Because mixing can take place more readily along surfaces of constant potential density than across them, and because lateral processes appear to be important in the formation and distribution of the Intermediate Water, the present study has been carried out by mapping various properties, including geostrophic flow, on two surfaces of constant potential density which, over large areas, correspond closely to these two tongues of low salinity. The patterns of distribution of temperature, salinity, oxygen, and phosphate along these surfaces are discussed in terms of the origin of the extreme characteristics and their modifications as the waters move.

Both of these surfaces intersect the sea surface in high southern latitudes. In the north the shallower density surface may occasionally intersect the sea surface off Kamchatka and the Kuril Islands, but the evidence available suggests that these occurrences are rare, brief, and limited to small areas. The deeper density surface never intersects the sea surface in the North Pacific, though it does lie nearer the surface there than in middle and low latitudes.

Between about 20°N and 40°N the minimum value of salinity in the water column is found near the shallower of these two density surfaces. This density surface does not effectively intersect the sea surface in the North Pacific: water of this density in the South Pacific is more saline. Therefore, the water at the salinity minimum between 20°N and 40°N does not originate at the sea surface in either the North or South Pacific. Its low salinity must be maintained by some other process.

In the equatorial Pacific a minimum in the vertical distribution of salinity is found on the deeper of the two density surfaces, which intersects the sea surface near Antarctica but nowhere in the North Pacific. The salinity on this surface near the equator is less than that above and below, but greater than that on this density surface farther north and south: the greater values near the equator must result from vertical exchange with the waters above and below.

On these two density surfaces the temperature and salinity are least in high latitudes. These properties are transmitted equatorward both by lateral mixing and by the general wind-driven circulation, which carries them equatorward in the eastern boundary currents and then westward in the equatorial currents. In lower latitudes the characteristic properties on these surfaces are altered by vertical exchange with the waters above and below, and both temperature and salinity are increased. It is emphasized that in these lower latitudes the two density surfaces lie several hundred meters below the mixed layer, and the vertical exchange occurs, not by convective overturn, but by processes of eddy conductivity and diffusivity.

In high southern latitudes, where these two density surfaces intersect the sea surface, the cooling, freshening, and aeration may occur by contact with the atmosphere. In the north, however, contact between the atmosphere and the deeper density surface is never established, and contact between the atmosphere and the shallower density surface is, at most, very limited, yet cooling, freshening, and aeration are observed on both density surfaces. It seems likely that this is a result of the same sort of vertical exchange that occurs in lower latitudes; that is, vertical eddy conductivity and diffusivity, without convective overturn to the depth of these two density surfaces.

It is proposed that the characteristics of the North Pacific Intermediate Water are formed in high latitudes by vertical mixing through the pycnocline. This mixing makes the waters in the pycnocline cold, low in salinity, and rich in oxygen in the subarctic region, and through lateral mixing (along density surfaces) and circulation these characteristics are transmitted equatorward from gyre to gyre. As a result, within the northern subtropical anticyclonic gyre the waters with density corresponding to that of the subarctic pycnocline are characterized by salinity less than that of the water above and below.

There is some reason to believe that vertical conductivity and diffusivity may also be effective in the formation of Intermediate Water in the South Pacific, though lateral mixing with surface water is undoubtedly much more important.

This system does not demand low-latitude upwelling of water from intermediate depths or sinking to such depths in high latitudes for balance of heat, salt, and water at intermediate depths, as would a thermal circulation.

It seems much more consonant with the known deeper circulation of the Pacific, which involves massive upwelling in the subarctic region rather than overturn to intermediate depths.

The distribution of dissolved oxygen is consonant with such a system. The concentration of oxygen on the two density surfaces is greatest in high latitudes, but the values in the south, where the density surfaces intersect the sea surface, are greater than those in the north, where the density surfaces have little or no direct contact with the atmosphere. From these high latitudes oxygen-rich Intermediate Water is mixed laterally into the subtropical anticyclones and carried equatorward in their eastern sectors. Consumption of oxygen reduces the concentration as the waters move. Replenishment of the oxygen at intermediate depths in the equatorial region by lateral processes involves transfer through the anticyclones, and the intermediate-depth waters that emerge from the equatorward edges of the anticyclones are much reduced in oxygen content. A steady state occurs only when the intermediate-depth oxygen near the equator has been severely reduced, and so strong a vertical gradient has been established that vertical diffusion, with whatever lateral replenishment there is, balances respiration and decay.

Contents

Introduction

The surface waters in the high latitudes of the Pacific are cool and the oxygen concentration is relatively high. In these latitudes precipitation exceeds evaporation and the surface salinity is relatively low. As a result the water masses lying in or near the mixed layer are relatively cool, low in salinity, and high in oxygen content at these extremes of their distributions. Because these characteristics appear to be spread throughout the middle and lower latitudes of the Pacific at intermediate depths (figs. 1–16), the water with these characteristics has come to be called intermediate water. To distinguish this intermediate water from other waters at intermediate depths that do not have these characteristics (as in the Antarctic and North Atlantic oceans), and to identify it where it lies near or in the upper mixed layer, the term will be written in this study with capitals: Intermediate Water.

This study began as an investigation of the northern salinity minimum, whose distribution in the region north of 20°N is shown in the Norpac *Atlas* (Norpac Committee, 1960a) by vertical sections along various parallels and meridians, and whose extension further south is indicated in fig. 3. Although the depth of the subsurface minimum varies from less than 300 to more than 800 m, the Norpac *Atlas* shows that over a region extending from about 20°N to 45°N (fig. 3) and from Japan nearly to America (fig. 8) the minimum lies (within the limits of the analysis) on or very near a surface of constant potential specific volume (approximated by thermosteric anomaly).

No such complete graphical description of the southern salinity minimum has been prepared, but near the equator it also lies very near a surface of constant potential specific volume (fig. 3). It corresponds in that region to the salinity minimum of the T-S diagram that Sverdrup *et al.* (1942) have used to describe Equatorial Pacific Water. They have shown that this water mass extends unchanged for the length of the equator and over a latitude range of 30° in the east and 10° in the west.

It was decided to approach the problem of the Intermediate Water by mapping the depth of these two surfaces of constant potential specific volume, distribution of temperature, salinity, oxygen, and phosphate-phosphorus along them, and, with a method suggested by Montgomery (1937), the geostrophic flow along these surfaces relative to suitable isobaric surfaces.

Particular terminology

Instead of writing "thermosteric anomaly" (Montgomery & Wooster, 1954), it will be convenient sometimes to write "δ_T"; the units will always be centiliters per ton (abbreviated cl/ton) and will not always be repeated. Surfaces of constant δ_T will be referred to as δ_T surfaces; the surface where δ_T equals 125 cl/ton, for example, will sometimes be designated the 125-cl/ton surface. "Lateral mixing" will be written for mixing along δ_T surfaces.

Acknowledgments

This study was supported principally by the Marine Life Research Program, which is the Scripps Institution of Oceanography's component of the California Cooperative Oceanic Fisheries Investigations, a project sponsored by the Marine Research Committee of the State of California. Additional support was provided by the Office of Naval Research. Some of the earlier part of the work was done at The Johns Hopkins University under National Science Foundation Research Grant G2935, "Analysis of Serial Oceanographic Observations."

The author wishes to give particular thanks to R. B. Montgomery for his support and encouragement throughout the several years this work has occupied. I also acknowledge with thanks discussions with J. P. Tully and R. S. Arthur.

1. Review of the literature

The literature has been roughly classified under several headings for convenience of presentation, although some of the studies deal with more than one of the headings and the headings themselves are not mutually exclusive. The early work of careful examination of few data against a limited background of theory is presented first, followed by more detailed descriptions and deductions, and then by discussions based upon particular processes or techniques. Studies that deal only with the South Pacific are presented in the final section.

1-1. Early speculations and hypotheses

Rumford (1798) put forward the hypothesis that since the temperature of maximum density of sea water is below the freezing point, the cooling of water in high latitudes causes it to become dense enough to sink all the way to the bottom and flow to the equator, where it rises to the surface, is warmed, and returns toward the poles. Humboldt (1831) claimed to have originated this hypothesis in 1813 (apparently he was unaware of Rumford's work), but he modified it when he noted that the eighteenth- and early nineteenth-century subsurface observations showed that in some high-latitude areas temperature increases with increasing depth, thus implying an increase in salinity with depth. He argued that if salinity played a significant part in the density variation of the ocean, then the simple thermal circulation might be modified or even reversed. Prestwich's (1875) comparison of the subsurface temperature distribution in the various oceans revealed that the North Pacific is consistently colder than the North Atlantic: he stated that the colder water of the North Pacific could not have come from the Arctic because the connection was too small; it could not have come from the surface waters of lower latitudes because they are too warm; and he proposed that it must therefore have come from the Antarctic Ocean by deep flow across the equator.

The *Challenger* expedition provided data that allowed the subsurface salinity distribution to be described and discussed and gave the first evidence of the salinity features (see fig. 3) that have been used to identify Intermediate Water. Buchanan (1876, p. 602), in discussing the *Challenger* expedition's measurements of salinity (which he tabulated as specific gravity at a common temperature, hence "dense" water signifies water of high salinity), wrote of the North and South Pacific oceans: "As a general rule, in both oceans between the parallels of 40°N and 40°S the specific gravity (reduced to 15.56 C) is greatest at or near the surface, and decreases more or less regularly until a minimum is reached, generally about 400 fathoms from the surface, when there is a slow rise, the bottom water being rather heavier. This general rule obtains in the Pacific and in the South Atlantic; but from the few observations obtained in the North Atlantic, there are indications of a departure from it."

Buchanan made this observation before the homeward voyage of the *Challenger* gave him an opportunity for a closer sampling of the North Atlantic. After the voyage he wrote (1877, pp. 81–82): "As in the Atlantic, the minimum specific gravity [salinity] is found usually at a depth of about 1000 fathoms [meters?], but in the Pacific we have light water approaching the equator from both sides, whilst in the Atlantic it occurs in a marked degree only from the south; and in the North Pacific by consequence the mean specific gravity is lower than in the south, the contrary being the case in the Atlantic. . . ." Still later, he published vertical sections along the *Challenger*'s track (1884, diagrams 5–7, 9–11) that revealed the general subsurface distribution of salinity (specific gravity) throughout much of the Pacific between about 40°N and 40°S.

In these findings of Prestwich and Buchanan the essential differences between the North Atlantic and North Pacific oceans are set forth: the North Pacific does not overturn to great depth, but receives its deeper waters from the south (and thus forms Intermediate Waters only); the North Atlantic does overturn to great depth, and no large quantity of low-salinity Intermediate Water is formed and transported equatorward.

Wüst (1929) recalculated the salinity from the specific gravity units used by Buchanan (1877) and with other limited data prepared a central and a western meridional section of salinity and temperature for the Pacific. He stated that the two tongues of low salinity extending equatorward from the cold high-latitude surface waters are better developed on the western section (though figs. 8 and 13 show that in middle latitudes this is not so) and that the northern tongue, though lower in salinity, is less extensive than the southern one. He proposed that there are intermediate-depth currents at the depths of the minima, a southern one from the Antarctic waters and another obviously originating in the marginal, cold Okhotsk Sea. He suggested that the (supposed) better development of the minima on the western section is a consequence of the earth's rotation, which forces the waters to move along the western edges of the ocean in their equatorward flow.

1-2.   More detailed descriptions

Studies based upon the shallow temperature minimum

Wüst's (1929) and many subsequent speculations about the origin of the Intermediate Water in the North Pacific associated it with the subsurface temperature minimum near 100 m revealed by the observations of the U.S.S. *Tuscarora* (Belknap, 1874) off the Kuril Islands, in the Bering Sea, and in the Gulf of Alaska. (The distribution of this feature can be inferred from figs. 2 and 28.) Makaroff (1894) found a similar temperature minimum near a depth of 100 m in the Okhotsk Sea. He concluded that it is formed by winter cooling of the mixed layer to a temperature less than that of the water immediately beneath, followed by summer warming of the upper part of the winter mixed layer. This conclusion implies that the water at the temperature minimum is, therefore, a relic of the winter mixed layer and that it indicates in summer the approximate conditions and deeper limit of the winter mixed layer. The presence of such a minimum precludes overturn to greater depth and the formation of deep water in the North Pacific.

Similar explanations of the subsurface temperature minimum were given by Lindenkohl (1897) for the Bering Sea and by Uda (1935) for the area off northeastern Japan; Uda was able to demonstrate the correctness of the explanation by means of direct measurements of temperature and salinity. Tully (1957) described the shallow temperature minimum in summer in the Gulf of Alaska and stated that it also appears to be a consequence of intense winter cooling of the mixed layer, followed by warming of the upper part of the column. Dodimead (1958, 1961) found that the area of the Gulf of Alaska covered in summer by the shallow temperature minimum varies from year to year. The shallow temperature minimum covered most of the Gulf of Alaska in the summers of 1955 and 1956 but was restricted to the northern and coastal areas in the summer of 1957. From the intervening winter data he showed that winter surface temperatures were above normal in the winter of 1956–1957, that the temperature at the minimum in the summer of 1957 was above normal, and that the area showing the shallow temperature minimum in summer was also much smaller.

A similar temperature minimum has been noted by Deacon (1937) in data from the high latitudes of the South Pacific. It is apparent from the Discovery Committee data (1944, stations 1662–99; 1947, stations 2158–2280; 1957, stations 2740–73) that over most of the area the minimum lies at a depth of approximately 100 m, but in the shallower water near Antarctica the U.S.S. *Glacier* (N.O.D.C. file 868, unpublished data) observed it to lie as deep as 450 m. Sverdrup *et al.* (1942, p. 609) were uncertain whether this summer minimum had been formed *in situ* during the preceding winter or by a northward flow of the colder stratum. Comparison of the values at the summer minimum with the available winter data (Discovery Committee, 1941, stations 879–978; 1957, stations 2817–41) suggests that the minimum is, like that in the North Pacific, and as Deacon (1937) had indicated, a relic of the winter mixed layer.

Uda (1935) noted the shallow temperature minimum off northeastern Japan in summer and argued that the presence of a deeper subsurface temperature minimum farther south (36°N–39°N) is evidence of a subsurface flow (Oyashio Cold Current) of this surface-formed water to the south, which penetrates the warmer waters of tropical origin and carries water of low enough salinity to appear as a salinity minimum after the temperature minimum is dissipated by vertical mixing. The data at that time were limited in vertical spacing and in horizontal extent and did not permit a clear discrimination between the levels and densities of the shallow temperature minimum (a relic of the winter mixed layer) and the salinity minimum that lies farther south in the pycnocline at a higher density and overlaps the horizontal extent of the temperature minimum in only a small area. He assumed, therefore, that they are coincident features and that water at the salinity minimum of the Intermediate Water has flowed directly from the winter-cooled mixed layer off northeastern Japan and in the Okhotsk Sea. Actually, the area off northeastern Japan quite often shows two temperature minima: besides the one near a depth of 100 m, which is a relic of the winter mixed layer, there is frequently a deeper, secondary minimum (at δ_T values from 80 to 130). The temperature and salinity are greater at the deeper minimum, and the oxygen concentration is very much less. The characteristics of the secondary minimum are probably derived from the colder deep water beneath the mixed layer in higher latitudes, possibly in the Okhotsk Sea, and the double minimum is probably a consequence of the upper waters of the Oyashio current flowing southward above the outflow from the Okhotsk Sea. Examples of the double temperature minimum can be found in the Norpac *Data* volume (Norpac Committee, 1960b) on *Tenyo Maru* stations 9, 13, 15, 16, 24, 33, 36, 37; *Ryofu Maru* stations 503, 506, 508, 521, 524; *Yushio Maru* stations 1007–9, 1012, 1015, 1024. The Transpac stations 50–52, 64, 66–70, 73, 77, and 78 also reveal the secondary minimum (Scripps Institution of Oceanography, 1965a).

Uda (1938, fig. 14) prepared a map to show the depth and salinity at the subsurface salinity minimum near Japan: north of about 40°N the minimum is at the surface and measures about 32.5 to 33.0 per mil. He also provided a table (1938, table 5) showing temperature and salinity from 21 stations where the temperature minimum is at or near a salinity minimum. Examination of this table, however, shows not only that the salinity at the salinity minimum is much greater than 33.0 per mil but that the density at these minima is in every case greater than can be produced by cooling water of 33.0 per mil to freezing temperature: the characteristics he shows at the salinity minimum cannot have been created at the surface in the region he indicated, nor can they have been the product of isopycnal mixing of other waters with such surface water.

Studies of the horizontal circulation

Koenuma (1936, 1937, 1939) attempted to account for the various water masses in the Pacific in terms of mixing and flow of various extreme water types. He proposed that the Intermediate Water is distributed by a cold current flowing southwestward beneath the Kuroshio (which he identified with Uda's Oyashio Cold Current), a west-southwestward countercurrent southeast of the Kuroshio, and, apparently, a larger-scale anticyclonic flow with its southern, westward-flowing limb south of 15°N. Sverdrup *et al.* (1942) stated that the Intermediate Water in the North Pacific is characterized by a salinity minimum with values from 33.8 per mil at a depth of 300 m north of 36°N to 34.1 per mil at about 800 m beneath the Kuroshio and that oxygen concentration at the minimum is greatest north of 36°N. They found the Intermediate Water present below the Central Water masses throughout the North Pacific and prepared, from the limited data available, a rough chart (fig. 202, p. 717) showing the salinity and depth at the minimum. On the basis of this chart they proposed a scheme of circulation of the Intermediate Water that shows it moving eastward from the area of formation and sinking off northeastern Japan, as specified by Uda (1935), and then moving southward and westward in a western-intensified anticyclone that occupies the western half of the subtropical Pacific. This is much like the scheme of Koenuma, except that no undercurrent beneath the Kuroshio is implied. Sverdrup and his associates were led to exclude the eastern area and thus to divide the subtropical anticyclone into two gyres, partly because of the double salinity minimum they noted off California at depths of about 300 and 550 m (fig. 8). It is the deeper of these that corresponds with the minimum to the west; on their chart they ignored the significant density difference and used values from the shallower one. The shallow minimum has been mapped in part by McGary & Stroup (1956), who suggested that it is formed by surface convergence in the west wind drift. This minimum is apparent in fig. 8 and has been charted in the Norpac *Atlas* (Norpac Committee, 1960a, plates 82, 83, and 90). It has been explained tentatively (Reid *et al.*, 1958) as a consequence of horizontal mixing of warm saline water from the central Pacific with cool, low-salinity waters of the California Current, in such a manner that the mixed layer values of salinity in the California Current are increased faster than those in the pycnocline, which remain low in salinity. This shallow salinity minimum continues beneath the California Current as it turns westward into the North Equatorial Current and is apparent near the equator in a vertical section at 160°W (fig. 3); in the Equapac data (Equapac Committee, in press) it is found as far west as 130°E.

A shallow salinity minimum that seems to correspond to that originating in the California Current (Reid *et al.*, 1958) can be detected in the Peru Current (fig. 13), according to the *Carnegie* data (Fleming *et al.*, 1945). This salinity minimum extends from the coast of South America, between 10°S and 36°S, to about 12°S 121°W; it apparently does not extend as far as the equator, as does the northern feature, nor does it have so great a westward extension. It was possibly this secondary salinity minimum of the Peru Current (fig. 13) that led Sverdrup *et al.* (1942) to postulate a separate eastern subtropical anticyclonic gyre in the South Pacific, an hypothesis that has not been accepted in the light of more recent data.

Although Sverdrup *et al.* (1942) did not discuss the density at the salinity minimum characterizing the Intermediate Water, they were apparently aware that surface data from the North Pacific did not reveal a density as great as that at the salinity minimum, for they carefully stated (p. 722) that the Intermediate Water "is probably formed in winter at the convergence between the Kuroshio Extension and the Oyashio and sinks from the surface in a manner similar to the sinking of the Antarctic Intermediate Water." In other words, they suggested that a mixture of saline Kuroshio water and cool Oyashio water might be further cooled in winter (at a time and in a place for which data were lacking) to the density observed at the salinity minimum.

Koshliakov (1961) has examined the geostrophic flow in the northwestern Pacific at various depths with respect to the 1,500-db surface, using data taken by the Japanese vessel *Mansu* from 1925 through 1928. He found that the direction of geostrophic flow did not vary with depth in the upper 800 meters and that the speed diminished only gradually with increasing depth. From this he argued that the flow of the Intermediate Waters is not markedly different from that of the surface waters and that both are part of a single wind-driven system.

Ichiye (1962) has considered the problem of whether the equatorward spread of the low-salinity waters of the North Pacific could be accounted for by consideration of meridional geostrophic flow. He calculated the geostrophic flow from the east-west pressure gradient and used as a reference the pressure surface that gives best agreement with the meridional Sverdrup transport calculated from wind stress. This allows calculation of meridional flow at all levels; he found that the calculated geostrophic flow at the level of the salinity minimum is in the wrong direction. He considered a model that also included lateral friction and concluded that, although it was perhaps better than the first model, it did not account for the subsurface equatorward extension of the waters of low salinity.

Vertical mixing and transport

The effect of vertical mixing and transport in forming the subsurface water masses has been discussed by various investigators. It is important to note that although the thermal structure of the subarctic region is characterized by a subsurface temperature minimum (in summer) and an underlying maximum (figs. 2 and 28) the salinity minimum is generally in the mixed layer or at the surface (figs. 3 and 29): the salinity increases monotonically with increasing depth. Mishima & Nishizawa (1955) have pointed out that in general the subsurface cold layer is not coincident with a salinity minimum but with salinity values as great as or greater than at the surface and have demonstrated this with data extending from northeastern Japan through the Bering Sea to Bristol Bay, crossing the Aleutian Island chain at 172°E. Their section is similar to the subarctic gyre sections (figs. 28 and 29) in the Oyashio, Northwest Pacific, and Bering Sea parts.

Tully & Barber (1960) and Dodimead (1961) have also shown that the shallow temperature minimum in summer in the Gulf of Alaska lies in an area where the minimum value of salinity is in the mixed layer or at the sea surface and that below the mixed layer (and, of course, below the temperature minimum) the salinity increases monotonically to depths of more than 2,000 m.

Masuzawa (1950) suggested that the deeper Intermediate Water may originate in the northern North Pacific from the upper deep water which near 50°N is warmer and more saline (figs. 2 and 3) than the upper waters. Saito (1952), in his study of the Oyashio Current, also discussed the shape of the T-S curves from most of the North Pacific. He classified the water of various areas by density and attempted to account for some distributions by both vertical and isopycnal mixing. In contrast to many previous and some subsequent investigators, he recognized in particular an Intermediate Water that lies beneath the "intercooled" water. He stated (pp. 107–8), "Then, it is reasonable to consider that the intermediate water in the Okhotsk Sea is formed by the mixture of the northern upper water with the North Pacific deep water."

Hirano (1957) discussed the subarctic water system to the north, as well as the area of the subsurface salinity minimum, and compared T-S diagrams from as far east as 160°W with those of the Bering Sea, the Okhotsk Sea, and the western subarctic region of the Pacific (excluding the Bering and Okhotsk seas). He proposed that the Central Water of the western subarctic might be formed by cooling and dilution of the warmer and more saline waters lying beneath the mixed layer in the Gulf of Alaska as they move westward south of the Aleutian Islands. He proposed further that this Central Water might in turn form the waters of the Bering and Okhotsk seas. He continued to maintain, however (1961), that the low-salinity Intermediate Water to the south is derived directly from the "intercooled" water (the subsurface temperature minimum left near 100 m by the summer surface warming), and proposed, as had Uda (1935), that the temperature-minimum feature disappears by diffusion, while the low salinity appears as a minimum to the south.

Koto & Fujii (1958) prepared a chart of the geostrophic circulation at the sea surface relative to the 800-db surface north of 45°N and west of 170°W and discussed the changes that take place in the temperature and salinity of the water flowing from the Gulf of Alaska northward into the Bering Sea and from there southwestward along the coast of Kamchatka and the Kuril Islands. They noted that as the waters pass through the Aleutian Island chain, vertical mixing decreases surface temperature (in summer) and increases surface salinity; in the deeper layers the salinity decreases by vertical mixing as the waters enter the Bering Sea and is further decreased as the waters move southwestward past the Kurils by mixing with the lower layers of the Okhotsk Sea. They stated that in the (cyclonic) "Western Subarctic Gyral," which is centered about 48°N and 165°E and corresponds to Hirano's (1957) "Western Subarctic Region," the salinity is greater than in the surrounding waters "on account of the ascending motion taking place in the gyral" (p. 170).

Lateral mixing

Kitamura (1958) recognized that the salinity minimum under the Kuroshio lies near the σ_t value of 26.8 (δ_T 126 cl/ton). Apparently this has been commonly known to Japanese oceanographers for some years, though it has not been possible to discover where the statement was first published. The results of the Norpac expedition (Norpac Committee, 1960a) show that this value holds for most of the North Pacific above 20°N, and Ichiye (1962) demonstrated it again for the western Pacific by averaging data from Japanese expeditions between 1933 and 1937.

Ichiye (1954, 1956) assumed, on the basis of the oxygen distribution, that the water beneath the shallow temperature minimum near a depth of 100 m is formed by sinking along σ_t surfaces from farther north. He plotted oxygen distributions on various σ_t surfaces in the area of the Kuroshio and found that at σ_t values of 26.8 and 27.0 (δ_T 126 and 106 cl/ton) oxygen values are relatively small and uniform in the southern Sea of Japan (west of 140°E) but increase markedly to the north and east on both sides of the Kuroshio. He also considered (1955) possible origins of the layer of minimum salinity lying beneath the Kuroshio. After examination of 36 vertical sections of geostrophic circulation, taken from observations made in 1951–1953, he concluded, as had Masuzawa (1950), that there is no countercurrent in the Intermediate Water beneath the Kuroshio, as was assumed by Uda (1935) and Koenuma (1936), and that the southwestward-flowing countercurrent at the southern edge of the Kuroshio (Koenuma, 1937) does not correspond to the salinity minimum on the vertical sections he studied. He concluded further that the salinity minimum must be a consequence of local lateral mixing along the conjunction of the Oyashio and Kuroshio. He proposed that large-scale lateral mixing occurs along the Oyashio frontal region and produces the Intermediate Water of the western North Pacific and denied that any of the Intermediate Water was transported southward by currents in the western Pacific.

Kuksa (1962) has reviewed much of the literature, particularly the Soviet investigations. After careful examination he also rejected the concept of a southward flow beneath the Kuroshio and proposed a different explanation of the observed distribution of the intermediate layer of low salinity in the middle latitudes of the North Pacific Ocean, as follows: the cold surface waters of low salinity formed in the Okhotsk and Bering seas and near the Kurils are carried southward by the Oyashio Current to the polar front (the junction of the Oyashio and Kuroshio); at the front they turn eastward and parallel to the Kuroshio extension; isopycnal mixing takes place in the stable layer between σ_t values 26.5 and 27.0 (δ_T 152 and 107) during this eastward flow; the waters of lower salinity from the north penetrate into the middle-latitude anticyclone by this lateral mixing with water of the same density and render the salinity in this density range lower than that at shallower and greater depths. He stated that it is unclear why the salinity minimum is not accompanied by a temperature minimum.

Kuksa (1962) did not discuss the particular origin of the cool, low-salinity water that is found in the southernmost part of the Oyashio, but in a later work (1963) he concluded that the origin of the salinity minimum of the Intermediate Water is the shallow temperature minimum. With this exception, his explanation is very much like that offered by Reid (1961a) and more fully presented here.

Frequency distribution of water characteristics

Cochrane (1958) has described the Intermediate Water of the Pacific Ocean in terms of the frequency distribution of water characteristics. He found a large quantity of water of 34.5 to 34.6 per mil, 4 C to 5.5 C, δ_T 60 to 80 cl/ton, at depths of 700 to 1,100 m near the equator (see figs. 3, 6, and 24), which he designated Tropical Water; he pointed out that the greater part of it lies beneath the salinity minimum, which he locates at 75 or more cl/ton. The water of δ_T 70–95 cl/ton and 3.5 C to 6 C, which merges with the Tropical Water, he called Southern Intermediate Water and stated that much of the Tropical Water appears to be only a modification of the Southern Intermediate Water produced by mixing with the saltier waters above and below. He stated that the Intermediate Water of the North Pacific is associated with the widespread salinity minimum at depths ranging from 300 to 800 m (see figs. 3 and 8) and that its range of characteristics is from 105 cl/ton to 135 cl/ton and from 5 C to 9 C. He stated (p. 124) that the frequency distribution of water characteristics seems consistent with the idea that this water is formed "in somewhat the same way as its southern counterpart. Water of characteristics near 2.0 C, 33.3 per mille descends from the surface and mixes with deeper, saltier waters reaching a smallest potential specific volume anomaly of 125 cl/t, possibly because of an increase in the rate of decrease of specific volume with depth."

The South Pacific

Deacon (1937) described the tongue of low salinity in the South Pacific as a mixture of Antarctic surface water and subtropical surface water sinking in the subantarctic zone (between the subtropical and antarctic divergences). He dealt mostly with the zonal flow but pointed out that in high latitudes the Intermediate Water "has a movement toward the east and is deflected toward the northwards in the eastern part of the ocean" (1937, p. 70). This is at variance with Wüst's (1929) description.

Sverdrup et al. (1942) stated that in the South Pacific the Intermediate Water originates at the surface in high latitudes from water of about 2.2 C, 33.8 per mil, and that it sinks and spreads northward between the surfaces where δ_T is about 69 and 88 cl/ton and mixes

with the waters above and below: the water mass thus formed is characterized by a salinity minimum that becomes less pronounced to the north. They stated that by the time that this water mass reaches 10°S it has gradually been transformed and that beneath the upper 200 m it conforms closely to a single temperature-salinity curve that extends all along the equator between 10°S and 15°N in the east and between 0° and 10°N in the west: they designate this as Equatorial Pacific Water.

Rochford (1960a and b) has discussed the distribution of salinity, oxygen, and phosphates on the surfaces where $\sigma_t = 26.80$ and 27.20 (δ_T 126 and 88 cl/ton) in the Tasman and Coral seas. The 27.20 surface was chosen as lying near or in the salinity minimum of the Intermediate Water from the south; the 26.80 surface was chosen as corresponding to the waters of the subtropical convergence, according to Deacon (1937). He classified the waters lying on these surfaces by their phosphate-salinity relation and found, on the 27.20 surface, high-phosphate, high-salinity water from the western equatorial area entering the Coral Sea through the Solomon Islands chain; moderate-phosphate, low-salinity water entering the Tasman Sea from the southeast and west of New Zealand; and low-phosphate, high-salinity water entering from the east, north of New Caledonia. On the 26.80 surface he found high-phosphate, moderate-salinity water entering from the east between New Zealand and New Caledonia; low-phosphate, high-salinity water (the dominant type on this surface) entering from the southwest; and water of low salinity and moderate phosphate content entering from the south.

Wyrtki (1962) has used the core-layer method of Wüst (1935) in describing and classifying by oxygen and salinity maxima and minima seven water masses in the southwestern Pacific (mostly the Tasman and Coral seas). On the basis of the salinity and oxygen distribution he inferred that the "Antarctic Intermediate Water" (the salinity minimum in the vertical plane) enters the Tasman Sea principally from the east, passing north and south of New Caledonia, having passed northward to the east of New Zealand; its inflow directly from the south, between Tasmania and New Zealand, he found to be very small, probably as a consequence of the generally southward flow of the East Australia Current. The shallower circulation he found to be similar to the surface circulation, i.e., generally counterclockwise with water entering from the northeast and east, turning southward along the coast of Australia and then eastward in the southern part of the Tasman Sea.

2. Plan of investigation

2-1. Lateral mixing and geostrophic flow

The importance of lateral mixing (along surfaces of constant potential density) in the ocean was proposed by Rossby (1936) and has been discussed by Montgomery (1939), Sverdrup (1939), and Sverdrup & Fleming (1941) in terms of mixing processes, and by Montgomery (1938) and Parr (1938), who examined distributions of various conservative and nonconservative properties on surfaces of constant σ_t in studying circulation in the Atlantic. The usefulness of such studies depends upon the assumption that flow and mixing take place along such surfaces to a much greater extent than across them and that movement of water from a particular source along such surfaces may form, in the maps of such quantities as salinity, temperature, oxygen, and phosphate, patterns that indicate the paths of flow. Flow patterns suggested by such maps can be verified and augmented in some cases by the geostrophic approximation of relative velocity. The interpretation of tongues of high or low concentrations as evidence of flow is in many cases sound, but it has been recognized (see Sverdrup *et al.*, 1942, pp. 503–7, for discussion) that the tongues need not represent the axes of flow in every case and that under some conditions the flow may parallel the isolines of a property.

It is hoped that by combining an investigation of the properties on surfaces along which they may mix most freely (surfaces of constant potential specific volume) with an investigation of the geostrophic flow along such surfaces (using a method suggested by Montgomery, 1937) a better understanding of the origin, flow, and decay of the characteristic water properties can be obtained.

2-2. Determination of surfaces of constant potential specific volume and calculation of geostrophic flow along such surfaces

Surfaces of constant potential specific volume have been approximated by surfaces of constant thermosteric anomaly. This quantity has been given the symbol δ_T (Montgomery & Wooster, 1954). It is the specific volume anomaly less the pressure-dependent terms, and it approximates closely the potential specific volume anomaly. In the two cases considered here the δ_T values are 125 cl/ton, which varies in depth in the Pacific from 0 to about 900 m, and 80 cl/ton, which varies from 0 to about 1,200 m. Raising the deepest observed water of these δ_T values adiabatically to the surface would reduce their δ_T values by less than 2 cl/ton: this is within measurement error, and therefore the δ_T value has been used as if it represented potential specific volume anomaly. Values of σ_t that correspond to the values of δ_T used on the figures are given in table 1.

The geostrophic flow along such surfaces has been computed from the gradient of the acceleration potential (Montgomery, 1937; Montgomery & Stroup, 1962) or Montgomery function, as it has been variously called. The function used here is

$$-\int_{p_2}^{p'} \delta\, dp + p'\delta'$$

where p is pressure, δ is specific volume anomaly, p' and δ' are the pressure and specific volume anomaly, respectively, at the depth of the surface along which the flow is to be computed, and p_2 is the isobaric surface to which the geostrophic flow is referred.

3. Preparation of vertical sections and maps

3-1.   Choice of materials

The materials used in this work (tables 2, 3, and 4) are from a wide range of sources: some were accumulated by the *Challenger* expedition nearly 100 years ago, and some are very recent. Choice among the materials was made on the basis of area, time, completeness, and quality. The Norpac, Eastropic, and Equapac expeditions each covered relatively large areas in short periods, and stations from these three were chosen in preference to other cruises that intersected their patterns. In other areas choice of the materials was based on season and need (most of the high-latitude stations were made in summer). During the several years that the preparation of the sections and maps has occupied, new data have become available; in some cases these were included; in others, the older materials seemed sufficient and were allowed to stand. It is therefore not possible to explain in detail each choice. Some differences appear in choice of stations on the two sets of maps. This is partly because of the difference in depth of the two δ_T surfaces. Because the deeper surface actually exceeded 1,000 m in depth in some areas, it was necessary to choose the 2,000-db surface as a reference in computing acceleration potential. Not all of the stations reached to 2,000 db: some stations were added to the deeper maps, but even with these the acceleration potential along the 80-cl/ton surface can be described only sparsely, particularly in the intertropical area.

3-2.   Treatment of data

The station data used on the various vertical sections and δ_T surfaces were plotted first on individual station graphs. Temperature was plotted against depth, and the other properties were usually drawn as characteristic curves against temperature, on a graph prepared and discussed by H. T. Klein (unpublished manuscript). Exceptions to the plot against temperature occurred in the case of high latitudes where temperature varied so little with depth that the characteristic curves would not have been useful, and in these cases all quantities were plotted against depth. After comparison of the various curves, and such redrawing as this comparison suggested, values of salinity, oxygen, phosphate, and depth were read for the various δ_T surfaces as well as any interpolated maxima or minima. The maps on δ_T surfaces were made directly from these readings. The vertical sections were made by plotting onto the section the observed values, the interpolated values at the δ_T levels of 40, 60, 80, 100, 125, 150, 180, 200, 250, 300, 350, 400, 450, and 500 cl/ton, and the interpolated maxima and minima. Many of the stations (particularly on the Norpac and Equapac expeditions) extended to little more than 1,000 m: in order to complete the sections data from nearby deep stations were used from 1,000 m down. In the case of stations lacking measurements of phosphate-phosphorus, it was also necessary on occasion to use phosphate data from nearby stations. Where this was done, the observed phosphate-phosphorus values were plotted below 1,000 m, but in the upper 1,000 m the values used were those interpolated at the various δ_T surfaces: they were plotted at the appropriate δ_T of the station used for the other properties. Such treatment is indicated where the vertical sections have short vertical lines at 0 and 1,000 m instead of the usual array of observation points.

On surfaces where δ_T is constant, the temperature is defined by the salinity. In table 5 are given the values of temperature that correspond to the salinity values contoured on the selected δ_T surfaces.

The vertical exaggeration of the sections is 5,550:1 in the upper 1,000 m and 1,110:1 below 1,000 m. The map is based on Lambert's azimuthal equal-area projection.

4. The surface where δ_T equals 125 cl/ton

4-1. Choice of the 125-cl/ton surface

The distribution of salinity in the North Pacific Ocean is characterized by a minimum value at the sea surface in high latitudes, and, beneath this surface minimum, by a monotonic increase with depth to a nearly constant value of about 34.68 per mil below 3,500 m (fig. 3). Beneath the surface salinity minimum in high latitudes the salinity increases rapidly in the pycnocline, and beneath the mixed layer in middle latitudes a tongue of low-salinity water is seen extending from high latitudes equatorward and downward, reaching along this section (160°W) a maximum depth of about 700 m near 25°N (a change of about 0.7 km in 2,000 km), and continuing as a recognizable minimum at about 600 m to about 10°N. In fig. 3 a shallower minimum is seen between the equator and about 15°N: the early data did not distinguish clearly between these two minima, and in some cases the shallower one was drawn as an equatorward extension of the deeper minimum (Wüst, 1929; Sverdrup, 1931). The two, however, are quite distinct from each other, as can be seen clearly in the eastern part of the Norpac Atlas (Norpac Committee, 1960a, plate 90) and throughout the Equapac data (Equapac Committee, in press) north of the equator.

It is apparent from the Norpac *Atlas* (Norpac Committee, 1960a, plates 82–90) that in the latitudes between about 20° and 45°, where the subsurface salinity minimum is best defined, the minimum lies very close to the surface on which δ_T is 125 cl/ton. The near coincidence of the salinity minimum with a particular value of δ_T can be determined only within the measurement error, and this includes the limitations imposed by the vertical spacing of samples, as well as errors in temperature and salinity. The indication that the salinity minimum lies at or near 125 cl/ton comes from consideration of the temperature-salinity curves. In the majority of the several hundred stations that

showed the minimum and, in particular, in those stations where samples were closely spaced near the salinity minimum, a smooth curve connecting the sample points shows the minimum near the value of 125 cl/ton, nearly always between the depths of the 120- and 130-cl/ton values. Some stations with sparse sampling near the salinity minimum could be drawn with the minimum at substantially different values of δ_T, but it was nearly always possible to draw the curves with the minimum at 125 cl/ton without introducing inflection points or other peculiarities. Significant variations in the δ_T value at the salinity minimum may become apparent when continuously recording instruments come into general use, but the present array of data in the area from 20°N to 45°N does not suggest any significant systematic variation. The value of δ_T that fits best to the salinity minimum over this area is about 125 cl/ton (obviously 123 or 127 would have done as well, considering the limitation in accuracy of salinity measurement alone.)

4-2. Distribution of properties on the 125-cl/ton surface

Depth

The depth of the surface where δ_T is 125 cl/ton is shown in fig. 17. No water of such high δ_T is seen south of 60°S, since the surface waters are too cold and saline there. In the north, however, where open-ocean values of 32.5 per mil are typical, water of this δ_T is not found any nearer the surface than about 140 m in these summer data. Winter temperatures are much lower, of course, but a salinity of at least about 33.28 per mil is required if a δ_T of 125 cl/ton is to be reached before the water cools to freezing temperature (Thompson, 1932). Over most of the region of low surface temperature the mixed-layer salinities are much less than 33.00 per mil (Sverdrup *et al.*, 1942; Norpac Committee, 1960a);

the δ_T of such waters cannot be reduced to 125 cl/ton by cooling. Recent maps of summer data in various years (Uda, 1963) show that north of 50°N the surface salinity was much less than 33.2 per mil except in a small area in the western Bering Sea, and there the values do not appear to have reached 33.30.

Examination of the comprehensive data and the analyses of salinity structure in the Gulf of Alaska (Tully & Barber, 1960; Dodimead, 1961; Uda, 1963; Dodimead *et al.*, 1963) show that the 125-cl/ton surface lies well beneath the mixed layer, even in winter.

Data from the central and northern parts of the Bering Sea in the winter of 1951 (Scripps Institution, 1963a) and 1955 (U. S. Navy Hydrographic Office, 1958; Scripps Institution, 1962) show surface values of δ_T greater than 125 cl/ton nearly everywhere: the exception is that a few of the stations in 1951 north of 64°N in the shallow water (less than 73 m), where temperature was everywhere within 0.05 C of −1.80 C, showed surface salinity greater than 33.28 per mil. Most of these high salinity values show unstable density structure, and the data may be invalid: the 1955 data in that area showed no surface salinity greater than 32.97.

In the Okhotsk Sea various summer expeditions show surface salinity less than 33.00 per mil (Agricultural Technology Association, 1954). On four winter cruises in February–May 1938 and February–May 1939 only 5 of 84 stations showed surface salinity above 33.13 per mil and only 2 reported as high as 33.28 per mil. One of these (Maritime Safety Agency, 1952, station 24 of cruise 60) on the shelf north of Sakhalin showed −1.44 C, 33.28 per mil, and 126 cl/ton at the surface. The other (Maritime Safety Board, 1962, station 10 of cruise 115) showed 1.10 C, 33.42 per mil, and 127 cl/ton. This salinity value is much greater than the surrounding ones (less than 33.13 per mil) and may be in error.

The small amount of more saline water entering from the Japan Sea might be expected to contribute to very dense water in the Okhotsk Sea. This water enters through the shallow (55 m) Soya Strait and flows eastward along the north coast of Hokkaido. Iida (1962) has presented data which may be taken to show that this water is quite warm (5 C to 19 C) as it enters. It is therefore low in density and can mix laterally with the less saline surface waters of the Okhotsk Sea before losing enough heat to the atmosphere to become very dense. Apparently this lateral mixing does occur, with the resulting mixture being fairly low in salinity and remaining in the mixed layer: Iida's (1962) data can be taken to indicate that the water from Soya Strait re-

mains well above the 125-cl/ton surface, and no evidence is found of cold saline surface water in the southern Okhotsk Sea. Akagawa (1958) has shown that in this region the salinity at the temperature minimum (about −1.6 C, near 100 m depth) is about 33.06 per mil. The δ_T of this water is about 143 cl/ton. This implies that water of δ_T 125 cl/ton does not intersect the sea surface in this region, and that the small amount of more saline water entering from Soya Strait does not substantially affect the salinity structure or the depth of overturn.

East of Hokkaido, Schott's chart of summer surface salinity (1935, *Tafel* XXVII) shows values of salinity less than 33.00 per mil, but since then various Japanese expeditions have observed some surface values as high as 33.96 per mil in both summer and winter (Japan Meteorological Agency, 1960a and b). These values are found in water that has come from the south, and even in winter the corresponding temperatures are too high for surface δ_T to be as low as 125 cl/ton except in the shallow inshore regions near Uchiura Bay (about 42.5°N 141°E) where the Central Meteorological Observatory (1952) reported values for January 1950 as high as 33.93 per mil in salinity and as low as 107 cl/ton in δ_T: these extreme values did not extend beyond the shelf at that time or in any of the other data reported by this agency.

Just east of Kamchatka and the Kurils the few (8 stations) winter data available (Maritime Safety Agency, 1951, p. 109; Agricultural Technology Association, 1954, p. 417) show no salinity values above 33.30 per mil and no δ_T below 139 cl/ton.

Over most of the North Pacific the data are adequate to show that values as low as 125 cl/ton never appear at the surface, but off Kamchatka and along the Kurils and Hokkaido they are not. Some information about winter mixed-layer conditions may be derived from the values at the summer temperature minimum in that area, but these data might not be representative: in summer the layer of minimum temperature may be warmed from both above and below, and the depth of the minimum may therefore be altered either upward or downward; Nansen bottle sampling may miss the extreme value; the colder deep water from the Okhotsk Sea may mix laterally with Pacific water to create deeper minima and thus add to the complexity.

Examination of the rather ample summer data shows that (except in the area just east of Kamchatka, along the Kurils, and just off Hokkaido) the surface where δ_T is 125 cl/ton lies well beneath the temperature minimum and at a depth where the oxygen concentration is substantially less than the saturation value for the corresponding temperature and salinity (Norpac Committee, 1960b; Hokkaido University, 1958–1963). In the critical area off Kamchatka, the Kurils, and Hokkaido, however, a few stations from particular cruises (Maritime Safety Agency, 1951, cruises 33 and 37 in 1935, cruise 38 in 1936; Hokkaido University, 1957) show values as low, or nearly as low, as 125 cl/ton at the temperature minimum. This finding suggests that at least in some years in a part of this critical area the surface where δ_T is 125 cl/ton may have intersected the sea surface. It seems probable that the occurrence of such low δ_T at the sea surface must have been limited to a small area and short duration: otherwise this occurrence and its subsequent effect would have been more obvious in the various data.

The maximum depth of the 125-cl/ton surface occurs in the western parts of the subtropical anticyclonic gyres and is about 800 m in the North Pacific and 700 m in the South Pacific. In general, the variation in depth of the surface corresponds to what might be expected from consideration of the known currents at the surface and the geostrophic approximation. The subarctic cyclone is represented by the shoaling near 50°N. The subantarctic cyclone (that flows around Antarctica) is indicated by the intersection with the sea surface. The subtropical anticyclones appear as great eccentric lenses. These lenses are separated from the equatorial trough by two zonal ridges that extend the entire breadth of the Pacific. It is this symmetry about the equator that first suggested the existence of a South Equatorial Countercurrent (Reid, 1959).

In the southern hemisphere the 125-cl/ton surface intersects the sea surface throughout the year. In these data (which include some observations from winter as well as summer) the latitude of intersection varies from about 46°S to 59°S, and the salinity at the intersection varies from about 33.90 to 34.35 per mil. Estimates based upon Schott's (1935) salinity map and these salinity data and Schott's temperature map suggest that in August the intersection is about 5° farther north, at salinity from 34.00 to 34.35 and at a temperature above 5 C.

Geostrophic flow

The geostrophic flow along the surface where δ_T is 125 cl/ton, with respect to the 1,000-db surface, is given by the gradient of the acceleration potential (fig. 18). The circulation suggested by the depth of this surface is more precisely defined. A poleward countercurrent (beneath the eastern boundary current at the surface) is indicated along the coast of North America,

less clearly indicated along South America. The eastward-flowing South Equatorial Countercurrent is indicated quite clearly on the various meridional lines of stations in the western half of the ocean between 3°S and 7°S. In the eastern area the data are sparse, but the eastward flow is fairly well documented between 5°S and 10°S and seems to be a more clearly defined feature at the depth of the 125-cl/ton surface than at the sea surface (Reid, 1959, 1961b; Wooster, 1961).

At the sea surface the California Current extends well into the eastern tropical Pacific before turning westward as the North Equatorial Current (Reid, 1961b). Along the 125-cl/ton surface, however, the current turns westward near 20°N, and the waters of the California Current do not flow into the eastern tropical Pacific. This leaves a large area of the northern eastern tropical Pacific that is not directly renewed by the equatorward-flowing Intermediate Waters. The corresponding area in the south is much smaller: this difference will be especially significant on the maps of oxygen distribution.

Salinity and temperature

The salinity and temperature on the 125-cl/ton surface (fig. 19) are least—about 33.4 per mil and 0.5 C—along the east coast of Kamchatka and in the Okhotsk Sea, where the 125-cl/ton surface lies about 200 m beneath the sea surface. In the South Pacific the least values on the 125-cl/ton surface are somewhat greater, about 33.9 per mil at about 5.2 C. From both of these extremes tongues of low salinity extend eastward, and the tongue in the South Pacific can be traced around the eastward and northward edges of the subtropical anticyclonic gyre all across the Pacific (note the 34.3 per mil value at 21°S in fig. 3) and into the Coral and Tasman seas.

Vertical mixing raises abruptly the salinity and temperature of part of the waters as they pass over the ridges and among the islands near New Caledonia. Part of the water passes northward through the Solomon Islands. As the other part turns southward, the vertical mixing with the warm, saline surface waters of the East Australia Current raises the salinity and temperature even further, and the greatest values on the 125-cl/ton surface—about 35.2 per mil and 11.8 C—are found near Tasmania. There is some suggestion of movement of this highly saline warm water from the Pacific to the Indian Ocean south of Australia. Taft (1963) shows water of δ_T 125 cl/ton with salinity greater than 35.0 per mil and temperature greater than 10.9 C lying southwest of Australia; this may be a continuation of the westward-flowing tongue south of Australia. Sverdrup *et al.* (1942, p. 615) have calculated that in the geostrophic transport above and relative to the 3,000-db surface there is a net westward flow immediately south of Australia.

It is perhaps surprising that the greatest values of salinity and temperature on this surface (near Tasmania) are found very near to the area of the least values on the 125-cl/ton surface in the southern hemisphere. These low values, where the 125-cl/ton surface intersects the sea surface at 50°S 110°–130°E, are about 33.9 per mil and 5.2 C. Downstream from there (along the 1.00 dynamic meter isopleth) the values increase to more than 34.3 per mil and 7.6 C at 160°W, probably as a consequence of lateral mixing with the highly saline waters brought south by the East Australia Current.

The tongue of low salinity reaching eastward along about 50°N does not extend equatorward, as does its southern counterpart, and between 50°N and the equator there is little east-west variation except near the coasts. Higher salinities and temperatures extend northward from 20°N in the countercurrent along the eastern boundary, and higher salinities are also seen in the western boundary current between 20°N and 35°N. Off central Japan the great eddy at about 36°N 145°E brings some cold, low-salinity water southward from the Oyashio.

Except for the Tasman Sea maximum, the highest values of temperature and salinity are found in a band extending along about 5°S from New Guinea almost to South America. The position of this band coincides roughly with the equatorward edge of the South Equatorial Countercurrent. Because the highest values are found near New Guinea, it might be proposed that the high values originate by vertical mixing in the shallow areas and passages of the Coral Sea, and that in the open Pacific the values are high as a consequence of eastward transport by the Countercurrent. This does not seem likely, however, except possibly quite near New Guinea. East of 180° the values of salinity and temperature at the maximum are nearly constant (almost within measurement error) at about 34.77 per mil and 9.9 C, and it seems unlikely that values could be maintained so constantly in such a narrow band over a distance of nearly 10,000 km by transport alone.

One might suppose that between the two high-latitude minima of salinity on this surface there should be higher values caused by large-area vertical mixing; since the original values at the minima are unequal and the waters approach the equator with unequal salinity,

the maximum need not occur precisely at the equator, but might be expected to occur south of the equator. In the region of the band the salinity and temperature are decreasing downward; it is therefore the shallower water that contributes the high salinity and temperature of the band. The shallower water has for its source of high values the very warm saline water in the pycnocline that extends equatorward from the southern subtropical anticyclone. That the higher values penetrate this surface here is probably a consequence of the shoaling of the 125-cl/ton surface at the east-west ridge at the southern edge of the South Equatorial Countercurrent. Nearer the surface the vertical diffusion may be intensified, and the water along this ridge may be made warmer and more saline as a result.

Dissolved oxygen

Where the 125-cl/ton surface intersects the sea surface in high southern latitudes, the concentration of dissolved oxygen is about the saturation concentration for the corresponding temperature and salinity. Values as high as 7 ml/l are observed near the intersection, and a great area of the 125-cl/ton surface shows values greater than 5 ml/l. In the northern hemisphere, however, where direct contact with the atmosphere apparently is not established, the maximum concentration observed is only a little more than 6 ml/l (about 0.8 of the saturation concentration for the observed temperature and salinity at atmospheric pressure). The areas of the 125-cl/ton surface in the northern hemisphere that show oxygen concentration exceeding 6 and 5 ml/l are equal to only small fractions of the corresponding areas in the south (fig. 20).

High values of oxygen extend around the southern subtropical anticyclone and into the Tasman and Coral seas. In the northern hemisphere the high values are extended eastward by the west wind drift and southward outside the countercurrent of the eastern boundary. A smaller tongue of high oxygen values can be seen extending southeastward in the Kuroshio-Oyashio eddy off Japan.

The tongue of high oxygen extending across the equator north of New Guinea may be evidence of cross-equatorial flow at this level. Sverdrup *et al.* (1942, p. 706) have suggested such flow on the basis of the T-S curves from the *Dana* data, and Wyrtki (1961) has discussed the flow using additional data.

Between the two areas of high oxygen concentration in the north and south the oxygen concentration is much less, but the distribution of oxygen in low latitudes is much more complex than is the salinity. The salinity distribution shows very little east-west variation near the equator, but there is a great difference between the concentration of oxygen at the eastern and western areas of the intertropical Pacific, and instead of a single extreme value located near the equator, there are at least two minima on this surface, north and south of the equator.

It is apparent from consideration of the vertical distribution of oxygen (fig. 4) that in the Pacific Ocean there is a source of oxygen in the northward-flowing deep and bottom water as well as at the sea surface. Processes of respiration and decay are likely to be most effective in reducing the oxygen concentration at those levels where replenishment by lateral diffusion and advection is least. At such levels a steady state occurs only when the oxygen has been severely reduced and a vertical gradient established that is so strong that vertical diffusion, with whatever lateral replenishment there is, will balance respiration and decay.

In both the North and the South Pacific the third source of higher-oxygen water to aerate the subsurface waters is associated with the Intermediate Water. It is most clearly seen in the South Pacific, in the great extension of the 4- and 5-ml/l isopleths northward at about 600 to 800 m and between 35°S and 20°S (fig. 4) and in the cross section at 27°S (fig. 14). The higher oxygen values lie somewhat shallower than the salinity minimum and have oxygen minima both above and below.

In the North Pacific the oxygen maximum of the Intermediate Water and the overlying minimum are less obvious but can be seen in the shape of the 4.5- and 5-ml/l isopleths between 20°N and 30°N at about 300 to 400 m depth (fig. 4) and in the cross section at 27°N by the 4.5-ml/l isopleth (fig. 9). The feature is less extreme in the North Pacific, probably because the Intermediate Water maximum is lower in concentration and lies much nearer to the surface water than in the South Pacific. Vertical diffusion of oxygen from both the surface and the Intermediate Water is more effective than in the South Pacific, where the vertical separation is greater.

Wüst (1935) suggested defining the Intermediate Water of the South Atlantic as that water lying between the two oxygen minima. Such a definition would be possible in the Pacific also, since there are two minima both north and south of the equator. However, a great part of the equatorial zone and some high-latitude areas do not seem to be covered well enough by both minima for the definition to be generally useful.

It is noteworthy that the deeper oxygen minimum north of 20°N lies near the 80-cl/ton surface, although south of the equator the oxygen concentration on the 80-cl/ton surface is relatively high (fig. 4). This emphasizes that the Intermediate Water takes part in the subtropical anticyclonic circulation and that it tends to lose its particular identifying characteristics in the equatorial region. Replenishment of the oxygen in the equatorial region by lateral processes involves transfer through the anticyclones, and the intermediate-depth waters that emerge from the equatorward edges of the anticyclones have been much reduced in oxygen content from the concentrations at the poleward edges. Thus, the oxygen concentration along the 80-cl/ton surface in the North Pacific cannot be maintained at a high level by mixing or transport from the rich 80-cl/ton waters south of the equator. In the close vicinity of the equator the oxygen concentration in the layer from about 100 to 300 m in depth is higher than that to the north and south: this has been associated with the presence of the Equatorial Undercurrent (Montgomery, 1954; Wooster & Cromwell, 1958; Knauss, 1960). Apparently this effect of the Undercurrent does not reach to the depth of the 125-cl/ton surface (about 400 m at the equator): the map of oxygen at this surface (fig. 20) shows a minimum value along much of the equator and a maximum value (perhaps) only near the Galápagos.

The conspicuous features of the intertropical region of the map of oxygen on the 125-cl/ton surface (fig. 20) are the areas of very low concentration. The northernmost of these is the largest and the easiest to account for in terms of the circulation. The minimum lies in the cyclonic gyre formed by parts of the North Equatorial Current and the North Equatorial Countercurrent. The waters of the California Current on the 125-cl/ton surface that reach a latitude of 20° have already lost a substantial part of their oxygen (as they have gained in temperature and salinity) by lateral mixing with the countercurrent of the eastern boundary, as well as by consumption. Instead of flowing directly into the eastern tropical Pacific, they turn westward to form the North Equatorial Current and are further depleted as they cross the Pacific. There is some renewal of oxygen in the far west by lateral mixing with waters from the South Pacific that have crossed the equator north of New Guinea. This crossing is much more apparent on the map of phosphate (fig. 21) and on the maps of oxygen and phosphate on the 80-cl/ton surface (figs. 25 and 26). As a result of this renewal the North Equatorial Countercurrent brings eastward water of relatively high oxygen. Values greater than 1 ml/l are found as far east as 120°W. There may also be replenishment of this tongue of high oxygen by vertical diffusion: the 125-cl/ton surface lies in the layer of minimum oxygen, but the vertical gradient (fig. 4) is very weak, and advection must play an important part.

From this point of view the oxygen minimum extending westward from Central America appears to be a consequence of minimum lateral replenishment. The waters that reach it have traveled to the furthest position from the areas of aeration; they have come by circuitous routes (if the trajectories based upon relative geostrophic flow can be accepted) and have been subjected continuously to processes of respiration and decay; part of the waters of high and moderate oxygen concentration has turned poleward again in the western boundary currents, and part has mixed laterally with the northern eastern boundary countercurrent. Vertical replenishment is limited by the high stability of the intertropical pycnoclines, and a balance is struck between use and replacement only when very low oxygen concentrations and strong vertical gradients are established.

This explanation ignores the possible effects of horizontal variations in rates of utilization of oxygen. It is possible that different rates of respiration and decay may be in part responsible for the variations in oxygen distribution along the 125-cl/ton surface, but it seems likely that they are secondary or consequent effects. The area of minimum oxygen concentration in the eastern tropical Pacific is also an area of very high inorganic phosphate-phosphorus concentration (fig. 5), and the overlying waters are very rich in biomass of zooplankton (Brandhorst, 1958; Reid, 1962b). Although it may be proposed that decaying matter from the great standing crop of the overlying waters causes the low oxygen and high phosphate concentration, it seems more likely that the relationship of cause and effect is the reverse: if one supposes that the waters reaching the eastern tropical Pacific have already been depleted of oxygen by decay that has produced inorganic phosphate-phosphorus, then the eastern tropical Pacific should be an area of high phosphate even if there were very little production immediately above. If, however, processes of vertical mixing, diffusion, and upwelling pass some of the phosphate to the surface, one might expect that in these tropical waters very high productivity might result, and the deeper oxygen content might be further depleted by decay of sinking matter. Equatorial upwelling (Cromwell, 1953; Reid, 1962b) does enrich the surface waters of the eastern tropical Pacific, and the variation in depth of the pycnocline (Cromwell, 1958) allows very high phosphate values to reach shallow depths near the boundary between the North Equatorial Current and the

North Equatorial Countercurrent: this is approximately the axis of the tongue of low oxygen.

The oxygen minimum just south of the equator on the 125-cl/ton surface does not correspond to the northern one in latitude or with respect to the currents. Although the data are not nearly so complete, especially in the area between 95°W and 135°W, it is certain that the minimum is smaller in extent than the northern one and that its axis lies nearer the equator. The northern minimum is centered at the boundary between the North Equatorial Current and the North Equatorial Countercurrent. There is only marginal evidence of a corresponding minimum (about 10°S between 110°W and 145°W) in the southern hemisphere. The major minimum in the south is centered on the boundary between the westward-flowing Equatorial Current and the South Equatorial Countercurrent east of 130°W, and on the Equatorial Current west of there.

There are certain notable differences between the properties on the 125-cl/ton surface north and south of the equator. One of these is that the replenishment of oxygen in high latitudes is more effective in the south (fig. 20) and the concentration of oxygen generally higher at all depths beneath the mixed layer (fig. 4). Another is that the North Equatorial Countercurrent receives some renewal of oxygen by lateral mixing in the far west and from there is marked by high oxygen values all across the Pacific: the South Equatorial Countercurrent begins in an area where lateral mixing is reducing the oxygen content of the southern waters and increasing that of the northern waters. The North Equatorial Countercurrent is seen to move eastward past 180° longitude as a local high in oxygen, though its value is only a little more than 2 ml/l. The South Equatorial Countercurrent moves at that longitude with a greater value of oxygen, but still with so much less than the concentrations in the westward-flowing water along 15°S that no minimum develops between these two currents (unless, perhaps, the somewhat doubtful feature near 10°S between 145°W and 115°W represents such a minimum). In the western and central area the equatorial countercurrents are seen to carry eastward water of oxygen content higher than that of the Equatorial Current that flows westward between them. In the eastern area the minimum in oxygen is found slightly to the south of the equator. It is conceivable that this is a consequence of the smaller transport of the South Equatorial Countercurrent in the eastern Pacific. The stronger North Equatorial Countercurrent supplies oxygen to the area just north of the equator:

if the supply from the south is much less, then the minimum may develop south of the equator between the maximum brought in by the North Equatorial Countercurrent and the generally high values of the South Pacific.

Other features that may contribute to the difference in size of the two areas of low oxygen are the shape of the coastline and (perhaps as a result) the different equatorward extent of the eastern boundary currents. The oxygen minimum north of the equator has been tentatively explained as a "dead" area by-passed by the California-North Equatorial Current. The eastern limb of the southern subtropical anticyclone carries water of 4 ml/l to 20°S near the eastern boundary (fig. 20). With this, as well as the South Equatorial Countercurrent, for lateral sources of oxygen, the area is better supplied than that in the north, and the area of concentration less than 0.25 ml/l is correspondingly smaller.

Phosphate-phosphorus

Of the various properties discussed, phosphate-phosphorus is apparently measured least accurately or with the least uniformity. Considerable variation occurs in measurements made by different ships in a common area. For this reason no attempt has been made to contour in as close detail as Kitamura (1958) has done in the area east of Japan and Rochford (1960a and b) has done in the Coral and Tasman seas. Details, therefore, may be misleading, but the general distribution so closely parallels that of the oxygen (except near the sea surface) that some confidence may be placed in the map (fig. 21).

Relatively high values of phosphate (more than 2 μg-at/l) are found at the sea surface near 60°S. These decrease to less than 2 in the southern subtropical anticyclone, but after that the trend is toward high values near the equator. High values (more than 2.5 μg-at/l) are found on the 125-cl/ton surface in the Gulf of Alaska and the Bering Sea: the tongue of low phosphate extending across the equator in the west is perhaps evidence of cross-equatorial flow. The highest values of phosphate are found in the eastern tropical Pacific, with the maximum of more than 3 μg-at/l north of the equator and corresponding to the area of low oxygen content. The maximum south of the equator is of smaller extent and less extreme, as is the southern oxygen minimum. The two maxima are separated by a tongue of low phosphate carried from the western Pacific by the North Equatorial Countercurrent. As on the oxygen map, the phosphate concentration in the South Equatorial Countercurrent does not seem to be extreme enough, or the flow strong enough, to give more than marginal evidence of a tongue extending eastward between 5°S and 10°S, and the resulting asymmetric pattern matches the oxygen map quite closely.

5. The surface where δ_T equals 80 cl/ton

5-1. Choice of the 80-cl/ton surface

The average salinity in the South Pacific is greater than that of the North Pacific, and the salinity minimum that extends equatorward from southern high latitudes is of correspondingly higher salinity and lower thermosteric anomaly value. The minimum clearly does not coincide everywhere with the 80-cl/ton surface: in high latitudes the values at the 80-cl/ton surface are small, but even smaller values are found at shallower depths, where δ_T is near 100 cl/ton. In the equatorial region, however, the minimum value appears to lie quite near the 80-cl/ton surface (within the limitations of the data, as described for the 125-cl/ton surface in sec. 4-1) and corresponds to the salinity minimum in the T-S curve that Sverdrup *et al.* (1942) have shown to extend the length of the equator in the Pacific. It is for this reason that the 80-cl/ton surface was chosen for examination of the Intermediate Water of the South Pacific.

5-2. Distribution of properties on the 80-cl/ton surface

Depth

The 80-cl/ton surface lies more than 350 m beneath the sea surface everywhere in the North Pacific: it intersects the sea surface south of 60°S in the South Pacific. The line of intersection must vary with season, but these data (fig. 22), which include both summer and winter observations, suggest that the 80-cl/ton surface intersects the sea surface throughout the year, at salinities and temperatures from slightly more than those of the freezing point (33.87 per mil and −1.845 C) to more than 34.20 per mil and 3 C. However, eastward of 160°W the intersection must lie quite near Antarctica: the surface salinity from 65°S to 70°S has been observed to be much less than 33.87 per mil, both by the *Ob* (IGY World Data Center A, 1961; stations 402–13) and the *Discovery* (Discovery Committee,

1947; stations 22, 44, 50, 55, 61, 68, and 74) in summer and the *Discovery* (Discovery Committee, 1941, stations 991 and 992; 1957, station 2835) in winter (see fig. 19). A slight lowering of the salinity between 70°S and Antarctica, or an increase in the temperature, might occasionally keep the 80-cl/ton surface entirely below the sea surface in that area.

The various effects of freezing, melting, and precipitation may produce rather wide and complex variations in the shape of the line of intersection in those areas where the surface salinity is near the critical value of 33.87 per mil. The area of δ_T less than 80 cl/ton centered near 65°S 160°W may sometimes connect with the southern area or may sometimes disappear.

There is a substantial variation of the depth of the surface (fig. 22), and the variation conforms in general to what might be expected from consideration of geostrophic balance of the known shallower currents. The steepest slopes are found in the western boundary currents and the Antarctic Circumpolar Current: the slopes in the equatorial system are much weaker. The subtropical anticyclones are indicated as large depressions in the western areas.

Geostrophic flow

In middle and high latitudes the acceleration potential at the 80-cl/ton surface with respect to the 2,000-db surface (fig. 23) indicates a circulation very similar to that seen on the 125-cl/ton surface. Between the Tropics the field of acceleration potential defined by the available data varies by less than 5 dynamic cm. In this area the Coriolis acceleration is much smaller than in higher latitudes and the flow is apparently weaker along the 80-cl/ton surface than along the 125-cl/ton surface; the horizontal pressure gradients are correspondingly less. Undoubtedly the scarcity of data reaching to 2,000 m, individual errors, and the systematic differences between the results of the various expeditions add much to the difficulty. The temperature and salinity beneath 1,000 m in depth vary so little horizontally between the Tropics that the error in salinity of 0.03 per mil that has been estimated for the *Carnegie* data (Fleming *et al.*, 1945) would have introduced relatively large errors in geopotential, and these data were not included on fig. 23. The salinities from the *Vityaz 26* expedition may be in error in the opposite sense, though this has not been discussed. These and other errors, individual and systematic, combined with the weakness of the gradients, render the intertropical region too confused for the contours to be accepted with much confidence.

This confusion is undoubtedly more severe in the deeper, more nearly homogeneous tropical waters below 1,000 m: the depth of the 80-cl/ton surface (fig. 22) is well defined, by these data, even near the equator, and the variation is seen to be similar to that of the depth of the 125-cl/ton surface.

Salinity and temperature

The salinity on the 80-cl/ton surface (fig. 24) is lowest in the southern hemisphere. In the southern high latitudes the minimum value in the vertical distribution of salinity is found at shallower depths than the 80-cl/ton surface, but between about 15°S or 10°S (25°S in the Tasman Sea) and about 10°N the minimum value of salinity is found very near to this surface.

The low-salinity water from the south moves northward and then westward around the subtropical anticyclone (fig. 23), extending westward near the Tropic as a tongue of low salinity. As it enters the Coral Sea over the irregular ridges and through the narrow passages on either side of New Caledonia, the salinity and temperature of part of the water rise to more than 34.5 per mil and 5.2 C by vertical mixing. Inside the Coral and Tasman seas the 80-cl/ton surface nearly coincides with the minimum salinity.

The geostrophic flow divides in the Coral Sea, part turning southward as the East Australia Current and part turning northward near the Solomon Island chain, and then westward. The southward branch rises in salinity along the coast of Australia and can be identified to the west of Tasmania as a tongue of high salinity (on this surface; low in the vertical plane) flowing westward toward the Indian Ocean. The northward branch of the flow also becomes warmer and more saline as it passes the Solomon Islands, probably as a consequence of vertical mixing with the more saline and warmer waters above and the more saline, slightly cooler, waters beneath. It enters the equatorial region at a temperature and salinity of about 5.3 C and 34.52 per mil.

Along the coast of South America from 15°S to 55°S

the highest values of salinity and temperature are observed near the coast. The data south of 32°S are very scanty, but they may serve as further evidence of the poleward flow that is at best marginally indicated by the geostrophic approximation (fig. 23).

In the north the effect of vertical mixing is made apparent by the low salinities and temperatures that are found in high latitudes. This surface does not intersect the sea surface anywhere in the north, its least depth being only a little less than 400 m near 50°N 170°E. The minimum salinity and temperature of a little less than 34.1 per mil and 1.9 C are found on the continental edge of the Okhotsk Sea at a depth of about 700 m. This freshening and cooling cannot have come from direct contact with the atmosphere, since the 80-cl/ton surface lies (beyond any doubt) well beneath the pycnocline and the shallow temperature minimum that appears in summer as a relic of the previous winter's cold mixed layer. The cooling and freshening, then, must be a result of vertical diffusion of heat and salt to the colder, fresher waters of the pycnocline.

The warmest and most saline water in the subarctic gyre is found at the extreme eastern end at the coast of Canada, where it is joined by the warm waters of the subsurface California Countercurrent that flow poleward along the coast of North America from south of the Tropic. From the coast of Canada the waters flowing westward with the Alaska Current and through and around the Aleutian Island chain and into the Bering Sea are freshened and cooled by vertical mixing, and reach their minimum temperature and salinities along the Kuril Islands: later, in the southward-flowing Oyashio, the salinity and temperature rise again. A tongue of low-salinity, cold water extends eastward with the west wind drift at about 50°N, but near Japan there is only a slight extension southward (to about 36°N) of the cold, fresh water of the Oyashio.

In the northwestern sector of the northern subtropical anticyclone, both salinity and temperature decrease downstream, particularly where the Kuroshio runs eastward south of the waters from the Oyashio, and the anticyclone waters are freshened and cooled. In the eastern and southern sectors temperature and salinity increase downstream.

In the predominantly zonal flow near the Tropics the isopleths of temperature and salinity extend east-west: in the zone between 5°S and 12°N in the west, 15°S and 20°N in the east, the salinity and temperature are relatively uniform, near the values of 34.55 per mil and 5.5 C that characterize the salinity minimum of the Equatorial Pacific T-S classification of Sverdrup *et al.* (1942). These investigators proposed (p. 706) that "The

Equatorial Water is probably formed off the coast of South America by gradual transformation of the Subantarctic Water . . . and spreads to the north and to the west. In the extreme western part of the Pacific the *Dana* data indicate that some of this water enters the North Pacific Ocean."

Cochrane (1958) has shown that the volume of Pacific water in the ranges of 5 C to 5.5 C, 34.5 to 34.6 per mil and δ_T from 71 to 83 cl/ton is 511×10^4 km³. Water of this salinity range but in the temperature ranges 4.5 C to 5 C and 4 C to 4.5 C, with δ_T from 65 to 78 and 60 to 73 cl/ton, totals 709 and $1{,}184 \times 10^4$ km³, however. Because most of the water in these three classes lies between the Tropics, he designated them as Tropical Water and pointed out that the greater part of the Tropical Water lies below the salinity minimum. On the basis of the smaller volumes of the surrounding classes, he stated (in agreement with Sverdrup *et al.,* 1942) that the Tropical Water in the North Pacific is apparently an intrusion from the South Pacific.

It seems likely from inspection of fig. 24, however, that waters of the North Pacific, as well as those from the south, may contribute substantially to the equatorial system.

Dissolved oxygen

In higher latitudes of both hemispheres the oxygen is renewed, in the south by contact with the atmosphere and in the north by contact with the fresher overlying waters. The original extreme values of high latitudes (fig. 25) can be traced as tongues extending along paths indicated by the geostrophic flow, at least through the eastward-flowing limbs of the subtropical anticyclones. Where the 80-cl/ton surface intersects the sea surface in the south, the original concentration of oxygen is at or near saturation, more than 7 ml/l. The northern counterpart shows evidence of some high-latitude replenishment of oxygen content, as well as lowering of salinity, in the Bering and Okhotsk seas, but at the level of the 80-cl/ton surface the oxygen rises only to a little more than 1 ml/l. The oxygen distribution is more asymmetrical about the equator than is the salinity, and it is seen that extreme values of oxygen from the south extend not only into the Tasman and Coral seas but can be traced further northwestward across the equator (a sort of flow that the geostrophic approximation cannot suggest). The high values extending northward past the Philippine Islands remain as high as 2 ml/l, even off southern Japan: this must be water from the South Pacific because no such high concentrations are achieved by the vertical mixing in the Bering and Okhotsk seas.

In most of the subsurface layers of the ocean the concentration of oxygen decreases downstream. This cannot be so everywhere, however, if a steady state is being maintained in a circulating system. In figs. 20 and 25 the waters moving southward past Tasmania are increasing in oxygen concentration, by vertical or lateral diffusion, or both. The waters moving westward south of the Aleutian Islands along the 80-cl/ton surface are increasing in oxygen concentration also, and since the oxygen minimum lies near this surface in that area, the source of the higher oxygen may be the waters above and below, as well as those immediately to the south. After passing through the Bering Sea and along the Kuril Islands, where the waters undergo a rise in oxygen corresponding to the observed fall in salinity and temperature, they encounter and join the water of oxygen greater than 1 ml/l flowing eastward along 40°N from the Kuroshio, the latter including water that has been replenished from the South Pacific. The oxygen concentration decreases downstream from there, and values less than 0.5 ml/l are found off the coast of North America. Except for the tongue of low oxygen extending westward along 15°N, these low values off North America are the lowest found on this surface, and after the waters turn southwestward at about 30°N the oxygen concentration rises again and continues to rise in the westward flow along 20°N. The oxygen minimum north of about 15°N lies near the 80-cl/ton surface (figs. 4 and 9): this replenishment must be caused by vertical diffusion from the richer layers above and below. The southern part of the North Equatorial Current passes beneath water of even lower oxygen concentration (fig. 4), and its oxygen concentration does not rise so rapidly: it remains lower in concentration than the northern part of the current as far west as 170°E.

This is an especially interesting feature in that any inferences about flow that were based only on the distribution of oxygen on this surface in the northeastern Pacific would require a curious system of currents: there are more tonguelike oxygen features than recognized currents. The extra feature is the very low oxygen concentration along 15°N in the eastern Pacific. This feature was seen in a more extreme form on the 125-cl/ton surface, and the tentative explanation was offered that this was an area of very poor lateral replenishment, with a low but steady concentration of oxygen maintained by vertical diffusion. It appears from comparison of figs. 4, 20, and 25 that the low values near 15°N on the 80-cl/ton surface are a consequence of the loss of oxygen upward to the even more impoverished level of the 125-cl/ton surface.

A smaller but comparable area of low oxygen concentration appears near 10°S in the east and nearer the equator in the west, separated from the northern lateral minimum by a tongue of high values extending along the equator. This minimum, like that to the north, is better developed on the shallower 125-cl/ton surface, and the low values here may likewise be partly the consequence of an upward loss of oxygen by vertical diffusion. That the values are higher in this minimum than in the north is probably a consequence of the generally higher concentration in the subsurface waters of the South Pacific.

Phosphate-phosphorus

In the South Pacific at high latitudes the values of phosphate-phosphorus on the 80-cl/ton surface (fig. 26) are slightly higher than those on the 125-cl/ton sur-

face. As with the oxygen, the distribution of PO_4-P suggests cross-equatorial flow in the western Pacific, and the path of the "newer" water (of higher oxygen and lower phosphate) can be traced eastward across the North Pacific along about 45°N. The PO_4-P concentration rises (as the oxygen falls) in the cross-equatorward flow and rises further as the waters turn eastward in the northern part of the subtropical anticyclone. As the waters turn westward in the southern part of the anticyclone (at about 20°N), the PO_4-P concentration decreases again, probably as a consequence of diffusion toward the shallower waters that have much smaller concentration. The tongue of high PO_4-P extending westward south of the Hawaiian Islands corresponds to the oxygen minimum that has been tentatively explained as a consequence of the extreme distribution along the 125-cl/ton surface.

6. The subarctic gyre

6-1. Choice of the section around the subarctic gyre

In the North Pacific the 125-cl/ton surface does not intersect the sea surface in any substantial area (section 4-2), and the 80-cl/ton surface does not intersect it at all, yet the effect of the sea-surface concentrations on the properties at the 125- and 80-cl/ton surfaces is marked. It is worthwhile, therefore, to examine step by step the modifications that take place in the waters flowing within the subarctic gyre.

Estimates of flow at the surface can be verified to some extent by direct observations of the set and drift of vessels; beneath the surface the quasi-geostrophic equilibrium of the ocean is the only known basis of estimation of speed and direction over large areas. While comparisons of surface geostrophic flow with averages of set and drift have been encouraging, the notable limitations in estimating trajectory (Reid, 1961b) must be kept in mind. That a surface gyre exists in the subarctic Pacific can be established from set and drift. The geostrophic flow at the surface relative to the 1,000-db surface repeats this pattern remarkably well, and the relative geostrophic flow will be used to estimate the deeper flow as well. The exact trajectory cannot be established, but a pattern based upon the relative geostrophic flow, bounded by land masses on the west, north, and east, and with zonal flow on the south, is probably not seriously misleading for the purpose of studying the change of properties along the flow.

The value of about 1.0 dynamic meter from the map of acceleration potential along the 125-cl/ton surface with respect to the 1,000-db surface (fig. 18) was taken as an indication of path (fig. 27), and vertical sections of the various properties along this path have been prepared (figs. 28–32).

6-2. Distribution of properties along the subarctic section

The location of the vertical section is shown in fig. 27. The first station (which is repeated as the last) is in the Gulf of Alaska. From left to right the section follows the flow westward south of the Aleutians, around the Aleutians and into the Bering Sea, around the center of the Bering Sea and southwestward into the extreme northwest Pacific, into and out of the Okhotsk Sea (although, of course, not all of the water in the subarctic gyre passes through the Okhotsk Sea), southward with the Oyashio, eastward with the west wind drift, and into the Gulf of Alaska and northward to the starting point.

Temperature

The section (fig. 28) shows clearly the subsurface temperature minimum that results from winter cooling. Its greatest value (about 4.9 C) is found in the Gulf of Alaska. Just south of the Aleutians the minimum value is less than 4 C and in the Bering Sea, less than 1 C. In the inner part of the Okhotsk Sea the minimum value is less than −1.5 C, and from there it increases in the Oyashio Current and west wind drift toward the greatest value of the minimum, observed in the Gulf of Alaska.

The depth of the temperature minimum varies from its shallowest (70–90 m) in the Okhotsk Sea and the northern Gulf of Alaska (where the surface salinity is very low) to its deepest (150–200 m) in the area northeast of Japan.

Beneath the minimum the temperature rises to a maximum at some depth below and then decreases downward. The greatest value at the maximum along this path is about 5.5 C and is found beneath the greatest value of the shallow temperature minimum, in the Gulf of Alaska, at a depth of about 150 m. (Still greater values of more than 7 C are found at the maximum to the south of this path.) Farther along the path, as the upper minimum becomes colder, the temperature at the maximum decreases and the depth of the maximum becomes greater. In the Bering Sea the maximum is about 3.5 C and is found at about 500 m. In the Okhotsk Sea the maximum is about 2.25 C and is found at an extreme depth of a little more than 1,000 m. Beyond the Okhotsk the temperature rises again, and the depth decreases toward the original values in the Gulf of Alaska.

Salinity

Nearly everywhere along the path salinity is low in the mixed layer, as a consequence of the high precipitation in those latitudes, and increases monotonically downward (fig. 29). The only exceptions are at the exit from the Okhotsk Sea and at the confluence of the Oyashio and Kuroshio, where two parcels of water of higher salinity appear near the surface. The most remarkable feature of the section is the intensive downward penetration of low-salinity water in the Okhotsk Sea.

Oxygen

Above the level of the temperature minimum the oxygen concentration (fig. 30) is large (almost everywhere greater than 6 ml/l), and along most of the path the maximum oxygen concentration is found in the summer thermocline (Reid, 1962a). Where low temperature and salinity values penetrate deep in the Okhotsk Sea, the oxygen value is correspondingly high. The oxygen minimum, which lies near 80 cl/ton over much of the North Pacific, lies much deeper along the northwestern part of this path.

6-3. Vertical distributions in the Okhotsk Sea

The most extreme penetration of the upper properties is apparent in the Okhotsk Sea. The temperature at the subsurface temperature minimum is less (below −1.5 C) than in the northwest Pacific, but the depth of the minimum value is nowhere much greater than 100 m. The temperature maximum below this lies between 800 and 1,000 m, in contrast to the Pacific maximum, which lies between 300 and 500 m in the northwestern area. The oxygen minimum in the Okhotsk Sea is less distinct than in the northwestern area of the Pacific and is about 1 ml/l greater; it also lies deeper, at or slightly below the temperature maximum. The general distribution of temperature, salinity, and oxygen may be taken to suggest that winter overturn extends only to about 100 m. At greater depths the waters, although originating in the open Pacific, are in restricted communication with it through the Kuril Island passages and probably have a longer residence time in the Okhotsk Sea than in comparable areas of the open Pacific. As a result, vertical diffusion (not overturn) has made the waters between 150 and 800 m substantially colder, less saline, and of higher oxygen content than those outside. Typical values at 500 m in the Okhotsk Sea are 1.2 C, 33.8 per mil, and 3.8 ml/l. Outside the Okhotsk Sea the values are 3.2 C, 34.2 per mil, and 0.7 ml/l.

7. Formation of the Intermediate Water

7-1. Modification of properties within the subarctic gyre

The 125-cl/ton isopleth (the upper dash line on figs. 28–31) represents, according to the assumptions (sec. 6-1) the trajectory of the flow along the subarctic section (and around the subarctic gyre).

Along the trajectory of the 125-cl/ton water the greatest values of temperature and salinity and the least values of oxygen (about 5.7 C, 33.88 per mil, and 1.9 ml/l) appear in the Gulf of Alaska. As the 125-cl/ton waters move westward from there, south of the Aleutian Islands (to the right of figs. 28–31), the temperature and salinity fall and the oxygen value rises. These changes continue along the path; in the eastern part of the Bering Sea (station TP30) the values are markedly different (about 3.4 C, 33.66 per mil, and 3.6 ml/l), and in the Okhotsk Sea the lowest temperature and salinity and the highest oxygen are attained (about 0.7 C, 33.41 per mil, and 5.5 ml/l). As the waters leave the Okhotsk Sea the trends are reversed, and in the Oyashio Current and the west wind drift the temperature and salinity rise and the oxygen value falls, and the original values are attained as the 125-cl/ton water reaches the northern part of the Gulf of Alaska.

Since it is evident (sec. 4-2) that convective overturn to the depth of the 125-cl/ton surface does not take place in the Gulf of Alaska, the North Central Pacific, or the southeastern Bering Sea, the cooling, freshening, and aeration of the 125-cl/ton water observed in that area must be accomplished without direct contact with the atmosphere. About half of the total modification of properties that takes place from the Gulf of Alaska to the Okhotsk Sea along this path has occurred by the time that the water reaches the central part of the Bering Sea (fig. 27, station TP30). This suggests that the cooling, freshening, and aeration of the 125-cl/ton

water that take place along this path in the northern part of the gyre cannot occur by convective overturn. Processes of vertical eddy conduction and diffusion at the depth of the 125-cl/ton surface must account (by upward transfer of heat and salt and downward transfer of oxygen) for the low values of temperature and salinity and the high values of oxygen that appear on this surface in the northern part of the subarctic gyre.

Only in a narrow band off Kamchatka, the Kuril Islands, and Hokkaido can values of δ_T near 125 cl/ton ever occur at the sea surface, and the available evidence (sec. 4-2) suggests that the occurrences must be rare, brief, or confined to a very small area. It seems likely, then, that the modification of properties on the 125-cl/ton surface in the subarctic gyre is usually accomplished (as it always is in the subtropical gyre) without direct contact with the atmosphere.

The trajectory represented on fig. 27 was based on the 125-cl/ton surface, but it can serve fairly well to represent the circulation on the 80-cl/ton surface (fig. 23). Water of 80 cl/ton never intersects the sea surface in the North Pacific, even in the critical area near Kamchatka and the Kuril Islands, but cooling, freshening, and aeration are seen to occur on the 80-cl/ton surface in the northern sector of the subarctic gyre. This must be a consequence of vertical diffusion.

Another area of the northern Pacific where aeration by vertical diffusion is apparent is at the head of the Gulf of California. The subsurface water at the mouth of the Gulf has properties like those offshore (a salinity minimum of about 34.55 per mil near 80 cl/ton and an oxygen minimum of about 0.10 ml/l, somewhat shallower). On the 125-cl/ton surface, which lies at about 350 m (well beneath the deepest penetration of the mixed layer), the salinity, temperature, and oxygen are increased by vertical transfer of salt, heat, and oxygen from about 34.60 per mil, 9 C, and 0.10 ml/l at the mouth of the Gulf to more than 34.76 per mil, 9.8 C, and 0.50 ml/l at the head (figs. 19 and 20). It is probably the long residence time and the extreme evaporation in the arid Gulf climate (Roden & Groves, 1959) that make the vertical diffusion so apparent there. The effect at the depth of the 125-cl/ton surface does not extend beyond the Gulf, however.

7-2. Lateral transfer and mixing between the subarctic and subtropical gyres

The net upward diffusion of salt and heat that occurs in the northern part of the subarctic gyre must be balanced by appropriate inflow at the bottom and outflow near the surface. The importance of transpycnocline mixing and transport in the subarctic region has been demonstrated by Tully & Barber (1960), who estimated the upward flow of deeper water required to balance the net precipitation in the Gulf of Alaska. They discussed the salinity structure in terms of two sources: the deep and bottom water (which enters from the South Pacific) and the product of the high precipitation over the subarctic region. These two sorts of water do not appear near the surface in their original form but as a mixture; Tully & Barber assumed that the downward mixing of fresh water extends to the bottom of the halocline, which they locate at a depth of about 200 m. They calculated that the 0.6 m of net annual precipitation in the region (Jacobs, 1951) must mix with about 20 m ± 10 m of water of salinity of 33.8 per mil (from below the halocline) to maintain the average salinity of the upper 200 m: this gives an upward transport of 20 m per year (± 10 m) in the subarctic region and implies that there is a net equatorward transport of the mixed water.

During its eastward passage in the west wind drift the 125-cl/ton water gains heat and salt and loses oxygen content (figs. 28–30). Some part of this change may be a consequence of vertical diffusion, but the lateral gradients are large, particularly in the western area (figs. 19 and 20), and lateral mixing is probably more important. The effect of the lateral mixing is to reduce the salinity and temperature and raise the oxygen concentration of the 125-cl/ton water in the adjacent subtropical gyre. By continued lateral mixing with the northern water, the water of this density in the subtropical gyre comes to be characterized by a conspicuous salinity minimum throughout the gyre (figs. 3 and 8). An oxygen maximum is less conspicuously present (figs. 4 and 9) at a slightly shallower depth over much of the gyre.

7-3. Comparison of the formation of the South Pacific and North Pacific Intermediate Water

It is noteworthy that the oxygen maximum of the North Pacific Intermediate Water (best seen on the 27°N section, fig. 9) lies slightly above the salinity minimum. If the Intermediate Water were formed by sinking of a particular water type from the surface, it might be expected that beneath the surface the extreme values of salinity and oxygen would lie at the same position, and that if sinking occurred along isopycnals they would lie on the same density surface. Since it is not proposed that the Intermediate Water of the North Pacific is formed by sinking from the surface but by mixing through the pycnocline, this difficulty does not arise. The boundary conditions (that is, surface and bottom concentrations) are different for salt and oxygen concentrations. Oxygen concentration above the pycnocline is nearly in equilibrium with the atmosphere, and it decreases toward the equator. Below the surface layer there is, over most of the North Pacific, a decrease downward to a minimum and then an increase to the bottom. Salinity, on the other hand, is least at the surface in high latitudes and greatest at the surface in middle latitudes, and at the bottom has a common intermediate value. Since both vertical and lateral mixing depend upon the gradients, it is not surprising that the intermediate-depth extremes of oxygen and salt concentration occur at different levels.

It is also apparent that in the South Pacific the oxygen maximum of the Intermediate Water lies at a slightly shallower depth than the salinity minimum. This is particularly obvious in the Tasman Sea (Wyrtki, 1962). Possibly this finding signifies that some part of the Intermediate Water of the South Pacific derives its characteristic properties beneath the surface in high latitudes from diffusion through the pycnocline, rather than from direct contact with the atmosphere as surface water.

Wüst (1929) has postulated an equatorward flow of Intermediate Water driven by high-latitude surface cooling and intensified along the western boundaries by the earth's rotation. The presence of the tongue of low salinity (figs. 3 and 24) extending anticyclonically around the southern subtropical gyre (fig. 23) seems to rule out the possibility that this flow is purely meridional and thermally driven. Indeed, the distribution of properties, including oxygen, in the Intermediate Water is consistent with the calculated geostrophic flow (figs. 18 and 23), which resembles the primarily wind-driven circulation at the sea surface and does not indicate a primarily thermohaline drive. Particularly in the North Pacific, the heat and salt exchange between low and high latitudes is clearly maintained without overturn or sinking to the depth of the Intermediate Water by means of lateral mixing and a pattern of flow which is like that of the wind-driven surface waters. The tongue of low salinity is seen to originate, not by flow from the surface, but by transpycnocline mixing in high latitudes, which gives the water of δ_T near 125 cl/ton characteristic properties that may appear as extremes when they are laterally mixed into the subtropical gyre. Salinity and oxygen do appear as extremes; temperature, because of the significantly different boundary conditions, does not.

Likewise, some part of the north-south exchange of heat and salt in the South Pacific may be largely effected without the actual sinking of water. It is clear from the distribution of salinity on the north-south vertical section (fig. 3) and on the 80-cl/ton surface

(fig. 24) that vertical diffusion accounts for the low salinity on the 80-cl/ton surface north of 20°N. It may be that vertical diffusion also contributes to the low salinity on the 80-cl/ton surface in the region from 55°S to 65°S.

The intense horizontal gradients and the steep (1:200) slope of the isopycnals near 55°S–65°S have been interpreted sometimes as direct convergence and sinking, although the same horizontal density gradient is used to calculate a geostrophically balanced Antarctic Circumpolar Current. Correspondingly, great density gradients exist at the edges of the Gulf Stream and the Kuroshio; they have been accepted as evidence of geostrophic balance in the flow rather than as evidence of the sinking of water. In the absence of specific measurements of sinking, there must remain the possibility of significant lateral mixing along isopycnals whose slope is geostrophically balanced, as well as the possibility of a thermohaline-driven flow.

8. Comparison of the formation and movement of the Atlantic and Pacific Intermediate Water

The major water mass deriving its characteristics in the North Atlantic is not Intermediate but Deep Water. In seeking for a counterpart to the subantarctic Intermediate Water, Wüst (1935) tentatively identified a water mass north of 45°N and west of 25°W as characterized by a salinity minimum between 500 and 1,000 m. This identification was based upon two stations that clearly indicate a salinity minimum of less than 34.9 per mil at 45°N–50°N and about 40°W. Further data and vertical sections in the area (Fuglister, 1960, p. 83; Dietrich, 1960, p. 94) confirm the minimum, but it appears (as Wüst suggested) rather irregular. In the absence of a direct analysis of the feature, it will be dismissed as small and confined to the extreme northwestern Atlantic, almost entirely obscured by the North Atlantic Deep Water, from whose properties its own deviate but slightly.

Continuing along the lines of his earlier (1929) work, Wüst (1935) has described the distribution of the salinity minimum core layer extending from the Antarctic in the Atlantic Ocean as a sinking and spreading of the denser surface water of high latitudes. He noted that the potential density in the core layer is substantially less in the area of sinking than at the extreme northern extent of the minimum, in contradiction to what had been assumed previously. He attempted to account for this change as a consequence of the vertical salinity gradient, which is stronger above than below the core, so that the time-change of salinity is more rapid above the core than below as the waters move northward, and the core appears at successively greater depths (potential density) to the northward. He noted also that this implies that the salinity core layer must lie somewhat deeper than the strongest intermediate flow. However, having chosen the core layer to study, his interpretation of flow was restricted to the tongues observed in the surface described by that feature. In the areas influenced by the subantarctic Intermediate Water but not characterized by a salinity minimum, his method cannot apply, and in the substantial areas of the core layer in which tongues are absent or not clearly defined, only doubtful inferences about the flow can be made.

Certainly Wüst's method has succeeded in indicating a northwestward transport along northern South America of water of southern origin, modified in the equatorial region. Whether there is also a system of zonal flows in the Intermediate Water in the tropical Atlantic could not be inferred from his use of the core-layer method and the available data, although Montgomery's study (1938) at shallower depths does find zonal flow.

The South Atlantic and South Pacific Intermediate Waters seem somewhat similar when viewed in mid-ocean meridional sections. The Atlantic profiles prepared by Wüst (1935, *Beilagen* XXIII–XXVIII) are characterized by a salinity minimum and an oxygen maximum extending northward from high latitudes of the South Atlantic Ocean. (In contrast to the Pacific, the depth of the oxygen maximum appears to coincide with the depth of the salinity minimum.) As these waters encounter no Intermediate Water from the North Atlantic, but instead the warm, highly saline water of the central North Atlantic, the salinity minimum can be traced much farther north than in the Pacific, to about 25°N in the western area. From the origin of the tongue to the equator the salinity at the minimum is nearly the same (34.0–34.6 per mil) as that of the Pacific, although at the equator the minimum value is about 34.50 per mil in the Atlantic as compared to about 34.55 per mil in the Pacific. In the equatorial regions of the two oceans the δ_T at the minimum is about the same (78–82 cl/ton). The oxygen concentration in the Intermediate Water is about the same in Wüst's central Atlantic section (*Beilage* XXV) as in the 160°W section in the Pacific, but the tongue of high oxygen concentration extending across the equator in the western Atlantic (Wüst's *Beilage* XI) is of slightly higher oxygen concentration (3–3.5 ml/l) than the Pacific counterpart (2.5–3.1 ml/l). This may be a consequence of the longer path that the Pacific water has traveled before crossing the equator.

Comparing the vertical section at 27°S with the corresponding section for the Atlantic (Wüst & Defant, 1936, *Beilage* VII, XXII, and XXXVII, at 27°S–30°S) one finds several notable differences. In the South Atlantic the water is generally colder in the upper kilometer (Reid, 1961c) but significantly warmer between 2 and 4 km. The salinity between 500 and 1,000 m is about the same in the Atlantic and Pacific sections, but in the surface layer the Atlantic values are higher. Below 1,500 m the Atlantic salinities are generally higher. The oxygen concentrations on the two sections are about the same down to the depth of the Intermediate Water oxygen maximum, but beneath that the Atlantic values are generally greater. In the upper kilometer the most marked differences occur near the continental boundaries. The presence of relatively warm, saline, and oxygen-poor water near the coasts in the Pacific is a consequence of the stronger poleward boundary currents (the eastern one being subsurface). As a result of these currents the Pacific Intermediate Water shows strong east-west gradients, while in the Atlantic the mid-ocean values extend almost unchanged to the coast.

Wüst has prepared maps (1935, *Beilagen* IX–XI) of the properties in the core layer of the subantarctic Intermediate Water, and he has proposed that the major flow and spreading (*Ausbreitung*) of cold, low-salinity, high-oxygen water is along the western boundary of the Atlantic and that this water extends northward along the coast well beyond the Falkland Current. Although this conjecture has been accepted by Defant (1941a) and by Sverdrup *et al.* (1942), there is reason to doubt it.

Taft's (1963) maps of the distribution of salinity and oxygen on various δ_T surfaces (actually potential thermosteric anomaly, but the differences can be ignored here) in the South Atlantic and South Pacific can be used to examine the real differences between the two oceans. The δ_T surfaces Taft examined are those of 60, 80, 100, and 125 cl/ton: the water at Wüst's core layer varies from 105 at the source in high latitudes to 63 cl/ton at the "zero" concentration (Wüst, 1935, p. 123). Taft's 125- and 100-cl/ton surfaces show a salinity and oxygen variation in the Atlantic similar to, but less marked than, that in the Pacific: the low-salinity, high-oxygen water extends equatorward across 35°S in the Atlantic in the eastern or central area of the ocean. On the 80- and 60-cl/ton surfaces there is no clearly defined equatorward tongue.

Wüst's values at the core layer show some evidence of the cold, low-salinity Falkland Current to about 40°S, but a clearer evidence is seen on the sections of temperature, salinity, and σ_t along 40°S–42°S and on the maps of temperature, salinity, and σ_t at 200 m (Wüst & Defant, 1936, *Beilagen* VI, XX, XXXV, XLVII, LXII, and LXXVII). Colder, lower salinity

water is found at 80 cl/ton near the western boundary near 40°S than is found in the central area. Taft's (1963) map is not detailed enough to show this relatively small but significant feature. Wüst's inference of equatorward flow near South America beyond 40°S, and beneath the Brazil Current, is based principally upon his map of temperature at the core layer of minimum salinity rather than upon salinity and oxygen. On this map a tongue of low temperature extends equatorward along the western boundary to 20°S: he inferred that this was a consequence of stronger equatorward flow or mixing. The significance of a temperature minimum along a surface defined by minimum salinity is not obvious. Some clue may be gathered by examination of the corresponding part of the Pacific. A similar temperature minimum would appear on a map at the salinity core layer in the southwestern Pacific, beneath the East Australia Current, as can be seen from the temperature and salinity sections along 27°S (figs. 12 and 13). In this case it is obvious that the temperature minimum would be a consequence of erosion of the salinity core layer by the more saline poleward-flowing water above, which causes the salinity minimum to appear at greater depths and thus at lower temperatures than in the areas to the east. The "cold tongue" in the Atlantic lies beneath the Brazil Current and may be the same sort of feature as in the Pacific. The inference of equatorward flow is therefore not the more likely explanation.

Both Defant (1941a) and Sverdrup *et al.* (1942) apparently accepted Wüst's inferred flow of the Intermediate Water and assumed that the subtropical anticyclonic gyre of the South Atlantic Ocean was limited to very shallow depths. Sverdrup *et al.* (1942, p. 628) stated that this gyre ". . . is a system of shallow currents because the entire circulation takes place above the Antarctic Intermediate Water or within the troposphere. . . ."

This is curious, since Defant's maps (1941a and b) of the relative and "absolute" topography of various pressure surfaces seem really to indicate a subtropical anticyclonic circulation along the 800-db surface as well as at the sea surface. Also, the maps of σ_t at various depths (Wüst & Defant, 1936, *Beilagen* LXXVII–LXXXI) and of the depth of various σ_t surfaces (Montgomery & Pollak, 1942) may be taken to imply that the subtropical anticyclone obtains throughout the depth range Wüst assigned to the Intermediate Water. Riley's (1951, pp. 46–56) calculations of transport also indicate the presence of a subtropical anticyclone in the South Atlantic extending at least to 1,200 m. Kirwan's (1963) more recent work also indicates such a feature.

Montgomery's (1938) study of the circulation in the

southern North Atlantic includes only one surface (σ_t 27.0 or δ_T 107 cl/ton) that is deep enough to allow a comparison with the 80- and 125-cl/ton surfaces in the corresponding area of the Pacific: he compared his results with Defant's (1936) study of the *Troposphäre*, which lies above the Intermediate Water. Although he calculated the acceleration potential for one meridional section, he did not make general use of it, usually inferring current axes from the shapes of the isopleths on the various maps.

Montgomery noted that on the 107-cl/ton surface the high salinities originate in the North Atlantic and the low salinities in the South Atlantic: this is demonstrated clearly by comparing Taft's (1963) map at 100 cl/ton with Montgomery's map. This might suggest that lateral mixing, almost alone, accounts for the low-latitude intermediate values. On the Pacific maps at 125 and 80 cl/ton the salinities are low in both high-latitude areas: the higher concentrations near the equator emphasize the effect of vertical diffusion.

The depth of the 107-cl/ton surface in the southern North Atlantic (Montgomery, 1938, chart 2) shows a minimum about 5°N to 10°N; this is somewhat like the 125-cl/ton map (fig. 17) in the Pacific, but is not defined except in the west. The salinity chart (Montgomery, 1938, chart 3) in the Atlantic is different, as noted above, but also shows a much more complex series of tongues of high and low salinity. The oxygen on the 107-cl/ton surface (Montgomery, 1938, chart 4) seems to suggest, however, the same sort of circulation as the Pacific 125-cl/ton map.

Tables

TABLE 1. Values of σ_t(g/l) that correspond to the values of δ_T(cl/ton) used on the figures.

δ_T	σ_t	δ_T	σ_t	δ_T	σ_t
25	27.862	125	26.807	350	24.440
30	27.809	150	26.543	400	23.915
40	27.714	180	26.227	450	23.392
60	27.493	200	26.017	500	22.868
80	27.281	250	25.491	550	22.345
100	27.070	300	24.965	600	21.823

TABLE 2.   The stations used in the vertical sections along 160°W longitude, 27°N latitude and 27°S latitude.

Expedition and/or ship	Stations	Dates	Remarks	Data source
		160°W (listed south to north)		
U.S.S. *Glacier*	1, OB16	23 Jan. 1962		N.O.D.C., file 868
U.S.S. *Burton Island*	13	29 Jan. 1962		N.O.D.C., file 867
Ob cruise 3	381–396	2–11 April 1958		Acad. Sci. U.S.S.R.; data taken from I.G.Y. W.D.C. A (1961)
Downwind (*Horizon*)	18–14	20 Nov.–1 Dec. 1957	used for phosphate-phosphorus	Scripps Inst. Oceanogr. (1965b)
Monsoon (*Argo*)	VII-14, VII-16, VII-22, VII-24	6–12 March 1961		Scripps Inst. Oceanogr. (unpublished data)
Lotus III	3	15 Sept. 1958		Serv. Hydr. de la Marine (1960)
Downwind (*Horizon*)	14	20 Nov. 1957	used only below 1,000 m	Scripps Inst. Oceanogr. (1965b)
Lotus III	1	14 Sept. 1958		Serv. Hydr. de la Marine (1960)
Carnegie cruise 7	91	27 March 1929	used only below 1,000 m	Fleming *et al.* (1945)
Equapac (*Hugh M. Smith*)	60–79	22 Sept.–5 Oct. 1956		Austin (1957)
Equapac (*Stranger*)	3, 6, 9, 13, 17, 21, 24	22–31 Aug. 1956	used only below 1,000 m	Scripps Inst. Oceanogr. (1957)

TABLE 2. (*Continued*)

Expedition and/or ship	Stations	Dates	Remarks	Data source
Carnegie cruise 7	139, 141–143	22 Sept.–9 Oct. 1929	used for phosphate-phosphorus only	Fleming *et al.* (1945)
Vityaz cruise 29	4098, 4102, 4104, 4108	28 Oct.–3 Nov. 1958	used for phosphate-phosphorus only	Institute of Oceanology (1961)
Norpac (*Hugh M. Smith*)	114	28 Aug. 1955		Norpac Committee (1960b)
Carnegie cruise 7	140	3 Oct. 1929	used only below 1,000 m	Fleming *et al.* (1945)
Norpac (*Hugh M. Smith*)	112–86	20–28 Aug. 1955		Norpac Committee (1960b)
Chinook (*Spencer F. Baird*)	2, 1, 4, 5	8–15 July 1956	used only below 1,000 m	Scripps Inst. Oceanogr. (1963b)
C.N.A.V. *Whitethroat*	95, 93, 90, 88, 86, 84	30 June–4 July 1962	used only below 1,000 m	Dodimead *et al.* (1962)
Norpac (*Brown Bear*)	57, 46, 45, 44, 42	29–30 Aug. 1955		Norpac Committee (1960b)

27° N (listed west to east)

Norpac (*Shumpu Maru*)	68, 71, 73, 77, 85	3–4 Sept. 1955		Norpac Committee (1960b)
Transpac (*Spencer F. Baird*)	96	24 Oct. 1953	used only below 1,000 m	Scripps Inst. Oceanogr. (1965a)
Norpac (*Satsuma*)	50–28	15–24 Aug. 1955		Norpac Committee (1960b)
Carnegie cruise 7	109	29 May 1929	used only below 1,000 m	Fleming *et al.* (1945)
Vityaz cruise 25	3625	6 July 1957		Institute of Oceanology (unpublished data obtained through W.D.C. B)
Vityaz cruise 29	4355	24 Feb. 1959	used only below 1,000 m	Institute of Oceanology (1961)
Vityaz cruise 26	3879	16 Feb. 1958		Institute of Oceanology (unpublished data obtained through W.D.C. B).
Vityaz cruise 29	4343	19 Feb. 1959		Institute of Oceanology (1961)
Vityaz cruise 29	4335	16 Feb. 1959		Institute of Oceanology (1961)
Chinook (*Spencer F. Baird*)	16	8 Aug. 1956	used only below 1,000 m	Scripps Inst. Oceanogr. (1963b)
Transpac (*Spencer F. Baird*)	130	15 Nov. 1953		Scripps Inst. Oceanogr. (1965a)
Carnegie cruise 7	141	5 Oct. 1929		Fleming *et al.* (1945)
U.S.S. *Bushnell*	p	26 Aug. 1934	used only below 1,000 m	Scripps Inst. Oceanogr. (1961a)
Norpac (*Stranger*)	101–106, 113–121	10–24 Aug. 1955		Norpac Committee (1960b)
Norpac (*Spencer F. Baird*)	131	25 Aug. 1955	used only below 1,000 m	Norpac Committee (1960b)

27° S (listed west to east)

Gascoyne	1, 2, 4, 6, 8, 10, 11, 13	2–4 Feb. 1960		C.S.I.R.O., 1962
Vityaz cruise 26	3844	22 Jan. 1958		Institute of Oceanology (unpublished data obtained through W.D.C. B)
Dana (*Dana*)	3624, 3626	10–13 Dec. 1928		Carlsberg Foundation (1937)
Vityaz cruise 26	3827	1 Jan. 1958		Institute of Oceanology (unpublished data obtained through W.D.C. B)
Monsoon (*Argo*)	VII-22, VII-24	11–12 March 1962		Scripps Inst. Oceanogr. (unpublished data)
Downwind (DW) (*Horizon*)	15	23 Nov. 1957		Scripps Inst. Oceanogr. (1965b)
Equapac (*Hugh M. Smith*)	26	26 Aug. 1956		Austin (1957)
Downwind (DW) (*Horizon*)	37, 36	7 Feb. 1958		Scripps Inst. Oceanogr. (1965b)
Ob cruise 3	430	7 May 1958		Acad. Sci. U.S.S.R.; data taken from I.G.Y. W.D.C. A (1961)
Downwind (DW) (*Horizon*)	35	1 Feb. 1958		Scripps Inst. Oceanogr. (1965b)
Ob cruise 3	433	11 May 1958		Acad. Sci. U.S.S.R.; data taken from I.G.Y W.D.C. A (1961)
Downwind (DW) (*Horizon*)	34	29 Jan. 1958		Scripps Inst. Oceanogr. (1965b)
Step-I (*Horizon*)	66–54, 52, 50, 49	17–26 Nov. 1960		Scripps Inst. Oceanogr. (1961b)

TABLE 3. (*Continued*)

Expedition and/or ship	Stations	Number of stations used					Dates	Data source
		Depth	S	O₂	PO₄-P	ΔD		
Tethys (*Stranger*)	*3, 7, 9–11, 13, 15–22, 24–27, 29–31, 34, 56*	0 0	0 0	0 0	0 0	0 22	18 June–16 Aug. 1960	Scripps Inst. Oceanogr. (unpublished data)
Transpac (*Spencer F. Baird*)	*4, 7, 9, 11, 12, 14, 17, 20, 23,* 24, 30, 35, 38, 40, 42, 44, 46, 47, 49, 52, 54, 57, 59, 61, 63, 66, 68, 70, 73a, 78, 80, 82, 84, 85, 92, 94, 96, 99, *100,* 101, 103, 106, *107, 111,* 112, 114, *115,* 117, *118,* 120, 122, 124, 127, 129, 132, *134,* 135	0 0	0 0	0 0	43 42	0 53	26 July–18 Nov. 1953	Scripps Inst. Oceanogr. (1965a)
Vermilion Sea (*Spencer F. Baird*)	*19, 29–31, 36, 46,* 47, 50, 62, 66, *67*	4 11	4 11	4 11	0 0	4 6	25 April–24 May 1959	Scripps Inst. Oceanogr. (1965e)
Vityaz cruise 25	3606, *3607,* 3608, *3609,* 3610, 3614, *3616, 3622,* 3623, 3625–27, 3629–33, 3639, 3642, 3644, 3646, 3648, 3651–53, 3655–65, 3667–3669, 3671, 3673, 3674, 3678, 3680, 3681, 3685, 3686, 3688–92, 3694, 3696, 3698–3700, 3703, 3705, 3707–09, 3711, 3715, 3719, 3722–24, 3726, *3727,* 3728–37, 3740, 3742, 3744, 3746, *3747,* 3748, 3750–53, 3756–59, 3761, 3763–66	35 72	35 76	35 75	66 86	35 86	1 July–7 Oct. 1957	Institute of Oceanology, (unpublished data obtained through W.D.C. B)
Vityaz cruise 26	3775–95, (3797, 3798,) 3801, 3802, (3803,) 3804, 3805–10, (3811,) 3812, 3814, 3816, 3818, 3820–40, 3842–81	76 71	73 74	75 73	88 82	71 64	13 Nov. 1957–18 Feb. 1958	Institute of Oceanology, (unpublished data obtained through W.D.C. B)
Vityaz cruise 27	3891, 3892, 3898–3900	5 0	5 0	5 0	5 5	0 0	28 March–4 April 1958	Institute of Oceanology, (unpublished data obtained through W.D.C. B)
Vityaz cruise 29	4317, 4318, 4320, 4321, 4323, 4325, 4327, 4329, 4331, 4333, 4335, 4337, 4339, 4341, 4343, 4345, 4347, 4349, 4351, 4353, 4355, 4357, 4359, 4361, 4362, 4364, 4366, 4368, 4369, 4371, 4373–85, (4386,) 4387–95	0 29	0 29	0 29	53 52	0 42	2 Feb.–11 March 1959	Institute of Oceanology (1961)
William Scoresby	597, 606, 609, 612, 616, 629, 638, 646, (652,) 653, 668, 671, 686–88, 694, 701, 703, 705, 707, 708, 711, 719, 722, 734–37	28 26	27 26	18 17	14 13	27 17	19 May–27 Aug. 1931	Discovery Committee (1949)
TOTALS		2148 1974	2137 2005	1919 1806	1519 1362	1927 840		

TABLE 4.   List of stations used in the vertical sections around the subarctic gyre.

Expedition and/or ship	Stations	Dates	Remarks	Data source
Norpac (*Brown Bear*)	23, 8, 44	10–29 Aug. 1955		Norpac Committee (1960b)
Brown Bear	32	15 Aug. 1958	used only below 1,000 m	Fleming *et al.* (1959)
Mukluk (*Horizon*)	10	8 Aug. 1957	used only below 1,000 m	Scripps Inst. Oceanogr. (1965b)
Norpac (*Oshoro Maru*) (OSH.)	35, 23, 12, 10	3–21 July 1955		Norpac Committee (1960b)
Leapfrog (*Stranger*)	9	6 Aug. 1961	used only below 1,000 m	Scripps Inst. Oceanogr. (unpublished data)
Norpac (H.M.C.S. *Ste. Therese*) (S.T.)	54, 74, 7, 11	30 July–28 Aug. 1955		Norpac Committee (1960b)
Gannet (GA.)	28	10 Aug. 1933		Barnes & Thompson (1938)
Chinook (*Spencer F. Baird*)	10	29 July 1956	used only below 1,000 m	Scripps Inst. Oceanogr. (1963b)
Transpac (TP.) (*Spencer F. Baird*)	38, 30, 42, 73, 72, 70, 59, 58, 52	15 Aug.–20 Sept. 1953	Stas. 70 and 59 used only below 1,000 m. Sta. 73 used for phosphate-phosphorus only	Scripps Inst. Oceanogr. (1965a)
121 (*Amataka Maru*) (AM.)	46	17 July 1953		Agri. Tech. Assoc. (1954)
117 (*Ryofu Maru*) (RY.)	92, 37, 23, 82, 18, 28, 38, 94	11–19 July 1942		Agri. Tech. Assoc. (1954)
Norpac (*Tenyo Maru*) (TE.)	6, 10	30 July–1 Aug. 1955		Norpac Committee (1960b)
Norpac (*Yushio Maru*) (YU.)	1003	12 Aug. 1955		Norpac Committee (1960b)
Norpac (*Ryofu Maru*) (RY.)	458	16 Aug. 1955		Norpac Committee (1960b)
Norpac (*Hugh H. Smith*)	34, 41, 79, 87	30 July–19 Aug. 1955		Norpac Committee (1960b)
Chinook (*Spencer F. Baird*)	7	19 July 1956	used only below 1,000 m	Scripps Inst. Oceanogr. (1963b)
Vityaz cruise 29	4066, 4112, 4388	15 Oct. 1958–10 March 1959	used for phosphate-phosphorus only	Institute of Oceanology (1961)

TABLE 5.   Values of temperature (to 0.01 C) that correspond to the isopleths of salinity drawn on the maps of the δ_T surfaces. (Freezing points are 33.87 per mil, -1.845 C on the 80-cl/ton surface, and 33.28 per mil, -1.810 C on the 125-cl/ton surface.) Calculated from LaFond (1951).

$\delta_T = 80$ cl/ton		$\delta_T = 125$ cl/ton			
S	T	S	T	S	T
per mil	C	per mil	C	per mil	C
33.90	-0.95	33.40	0.59	34.40	8.14
34.00	0.72	33.50	1.82	34.50	8.62
34.10	1.88	33.60	2.82	34.60	9.11
34.15	2.39	33.70	3.68	34.70	9.52
34.20	2.86	33.80	4.46	34.75	9.82
34.25	3.30	33.90	5.16	34.80	10.05
34.30	3.71	34.00	5.82	35.00	10.93
34.40	4.47	34.20	7.02	35.20	11.76
34.50	5.16	34.30	7.58		
34.55	5.50				

Figures

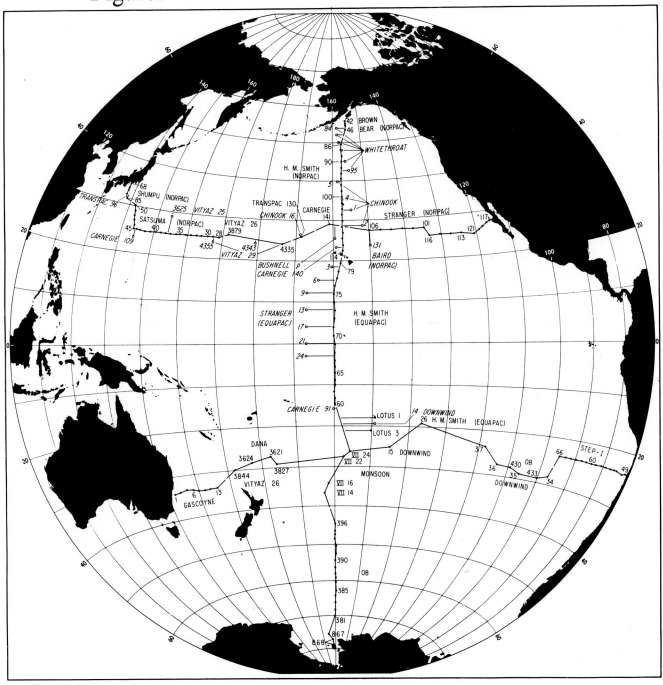

Figure 1.   Positions of the stations used on the vertical sections in figs. 2–16. Stations used only at depths below 1,000 m are indicated by open circles. The manner of preparation of the sections is described briefly in sec. 3-2. Data sources are given in table 2. (Lambert's azimuthal equal-area projection is used on this map and on figs. 17–27.)

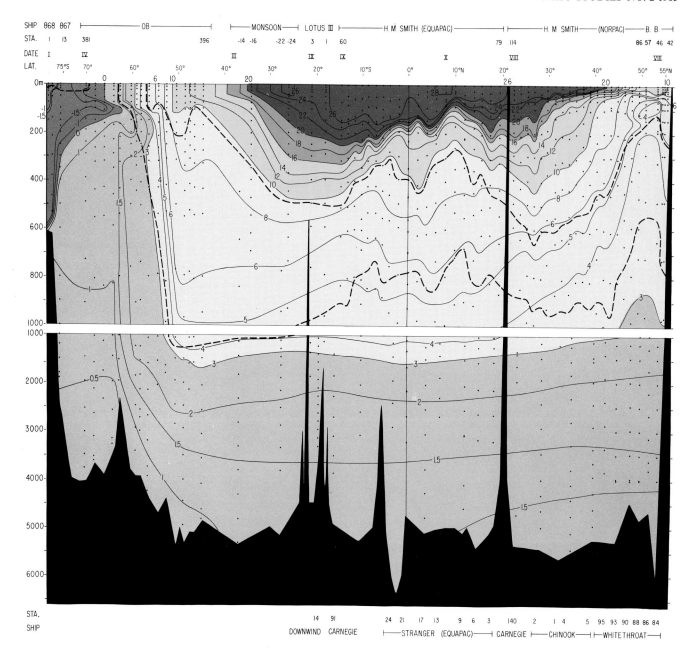

Figure 2.   Temperature (Celsius) along approximately 160°W from Antarctica to Alaska. The heavy dash lines on this and the other vertical sections indicate the depths of the 125-cl/ton and 80-cl/ton isopleths. Vertical exaggeration on this and all other vertical sections is 5.55×10^3 in the upper 1,000 m, 1.11×10^3 below 1,000 m.

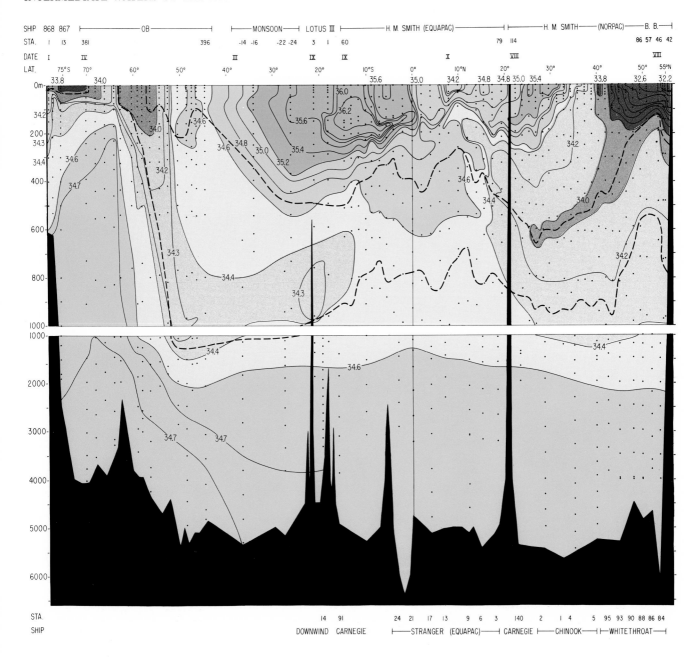

Figure 3.   Salinity (per mil) along approximately 160°W from Antarctica to Alaska.

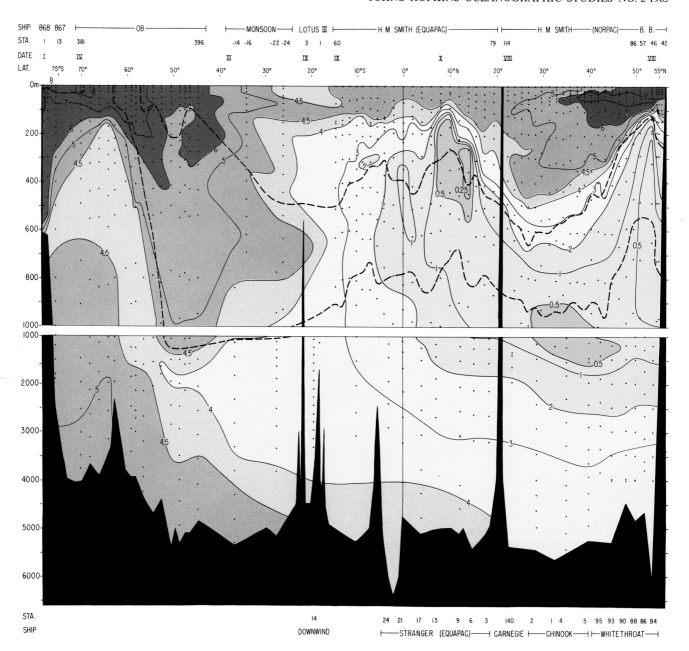

Figure 4.   Oxygen (in milliliters per liter) along approximately 160°W from Antarctica to Alaska.

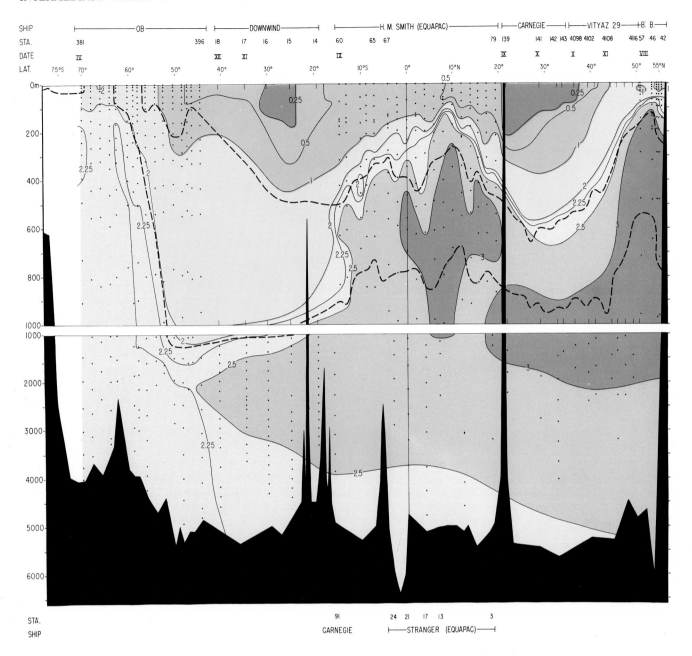

Figure 5.   Inorganic phosphate-phosphorus (in microgram-atoms per liter) along approximately 160°W from Antarctica to Alaska.

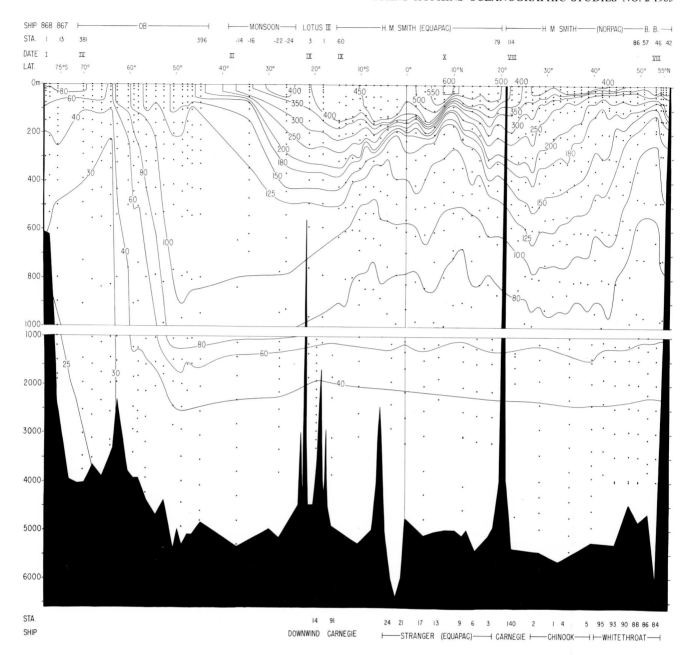

Figure 6.   Thermosteric anomaly (in centiliters per ton) along approximately 160°W from Antarctica to Alaska. Corresponding σ_t values are listed in table 2.

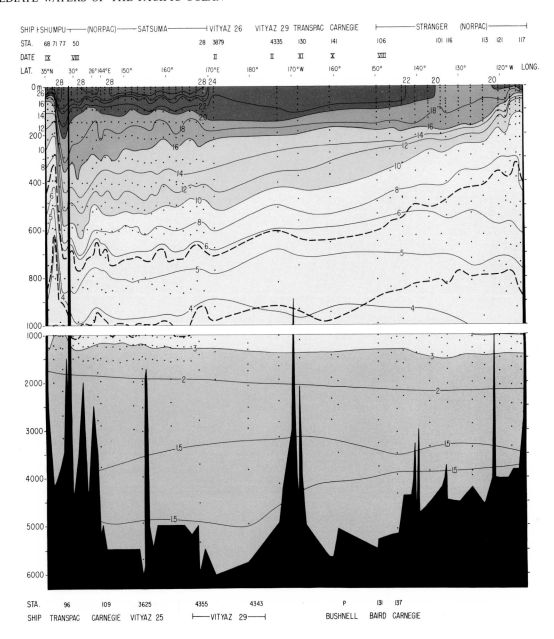

Figure 7.   Temperature (Celsius) along approximately 27°N from Japan to North America.

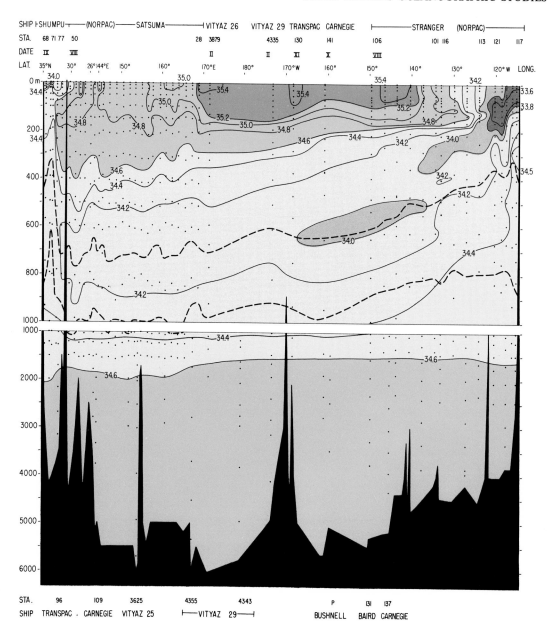

Figure 8.   Salinity (per mil) along approximately 27°N from Japan to North America.

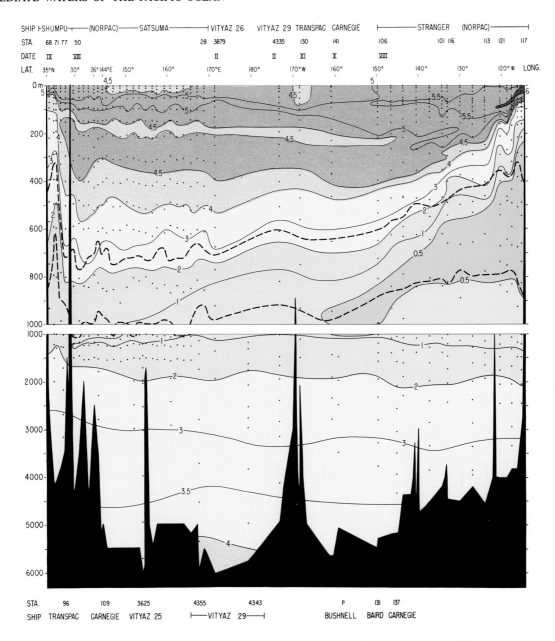

Figure 9.   Oxygen (in milliliters per liter) along approximately 27°N from Japan to North America.

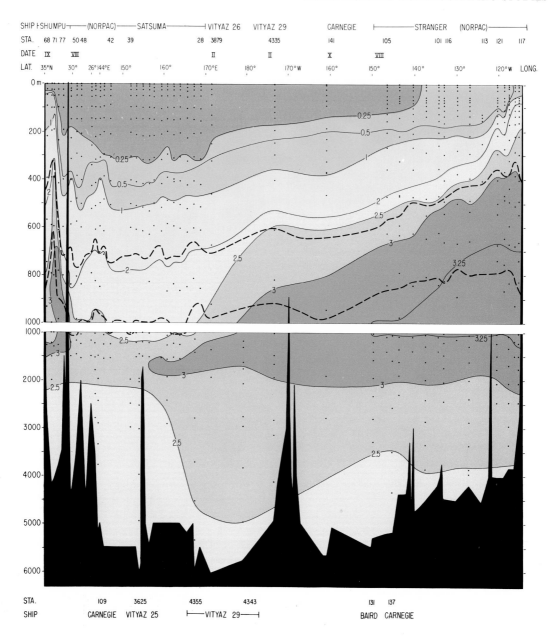

Figure 10.   Inorganic phosphate-phosphorus (in microgram-atoms per liter) along approximately 27°N from Japan to North America.

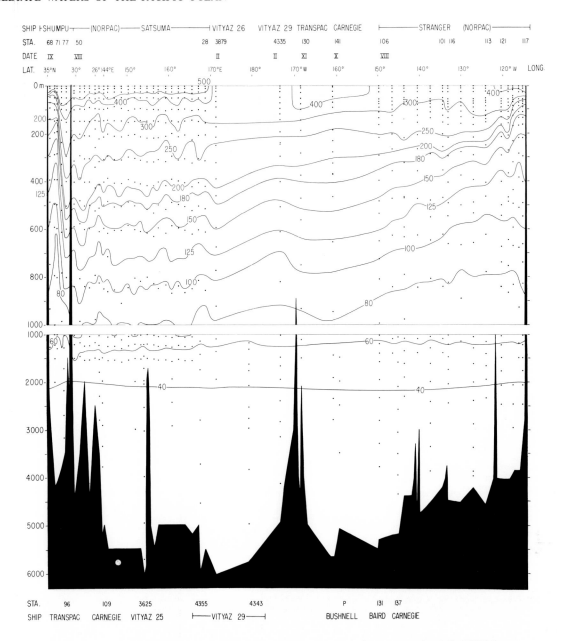

Figure 11. Thermosteric anomaly (in centiliters per ton) along approximately 27°N from Japan to North America.

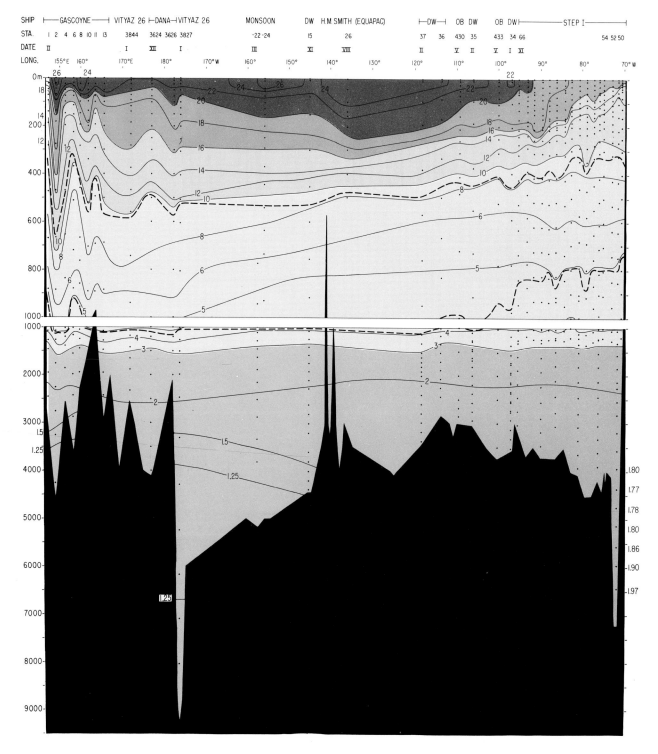

Figure 12.   Temperature (Celsius) along approximately 27°S from Australia to South America.

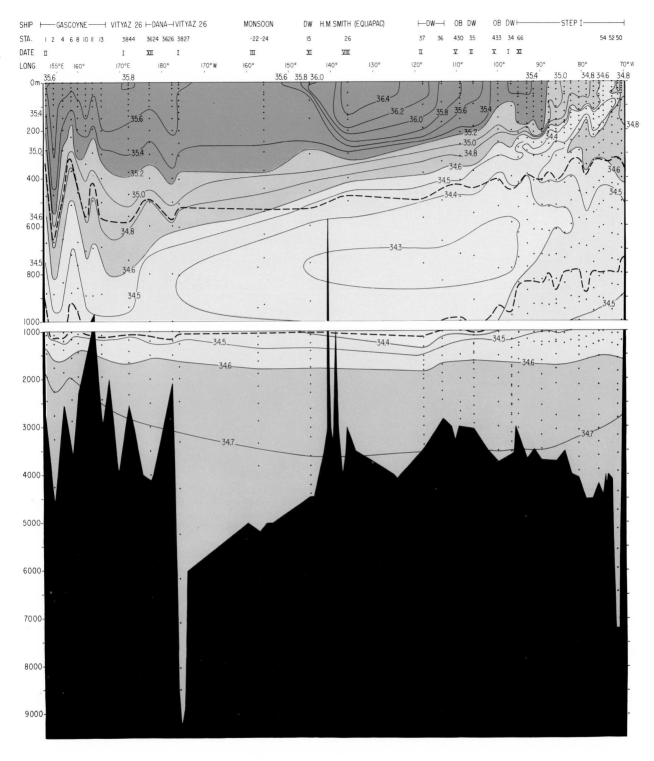

Figure 13.   Salinity (per mil) along approximately 27°S from Australia to South America.

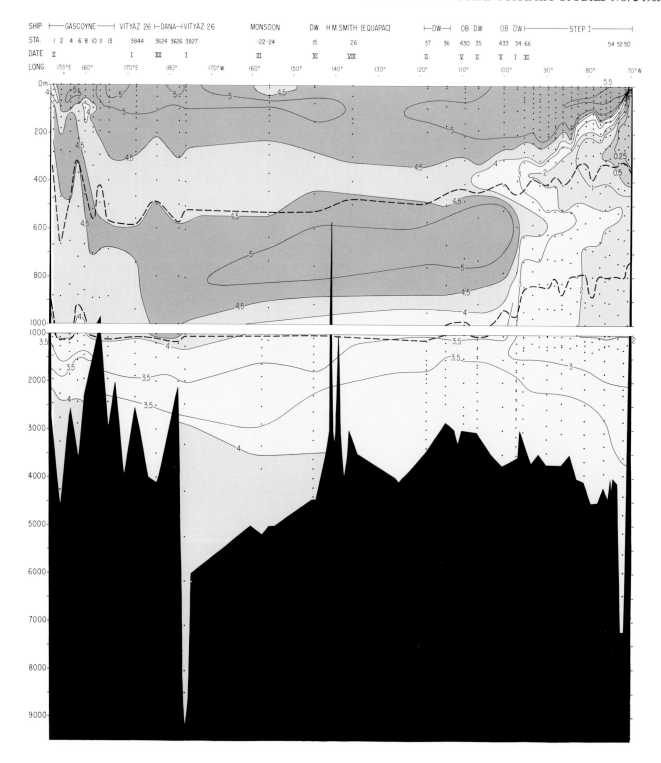

Figure 14.   Oxygen (in milliliters per liter) along approximately 27°S from Australia to South America.

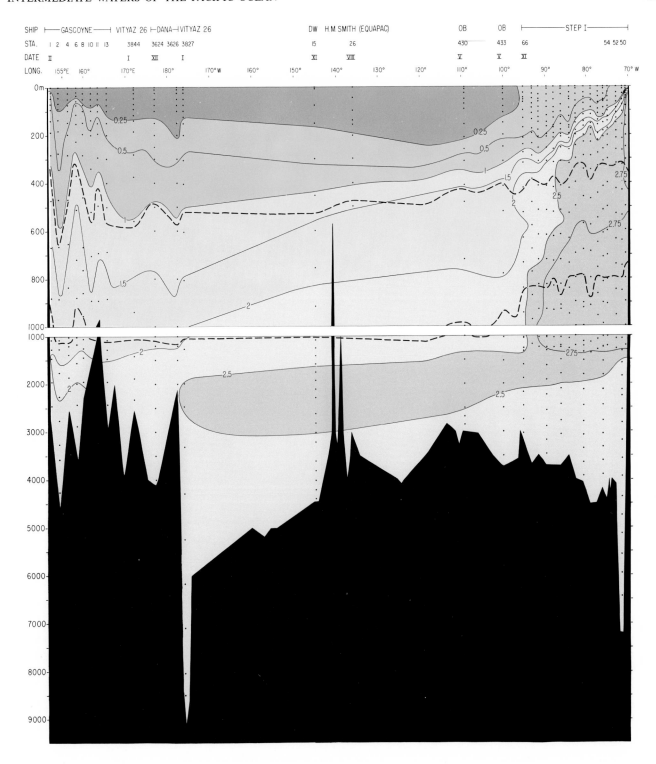

Figure 15.   Inorganic phosphate-phosphorus (in microgram-atoms per liter) along approximately 27°S from Australia to South America.

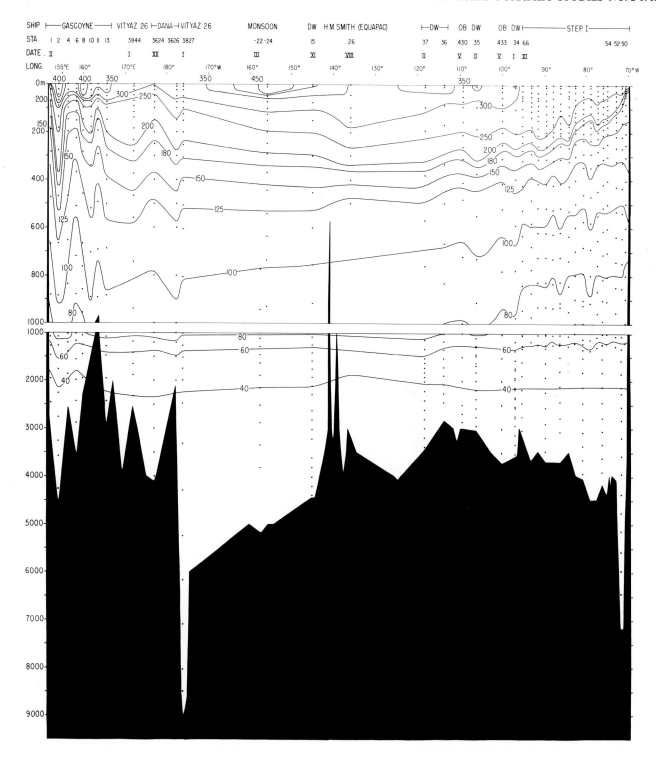

Figure 16.   Thermosteric anomaly (in centiliters per ton) along approximately 27°S from Australia to South America.

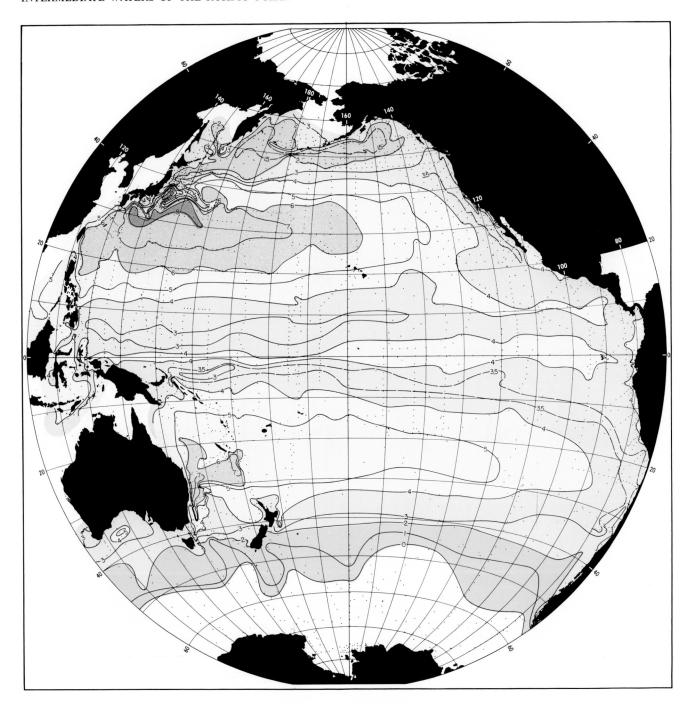

Figure 17.   Depth (in hectometers) of the surface where $\delta_T = 125$ cl/ton. The manner of preparation of the maps at 125 cl/ton is described briefly in sec. 3-2. The data sources for all the maps are given in table 3.

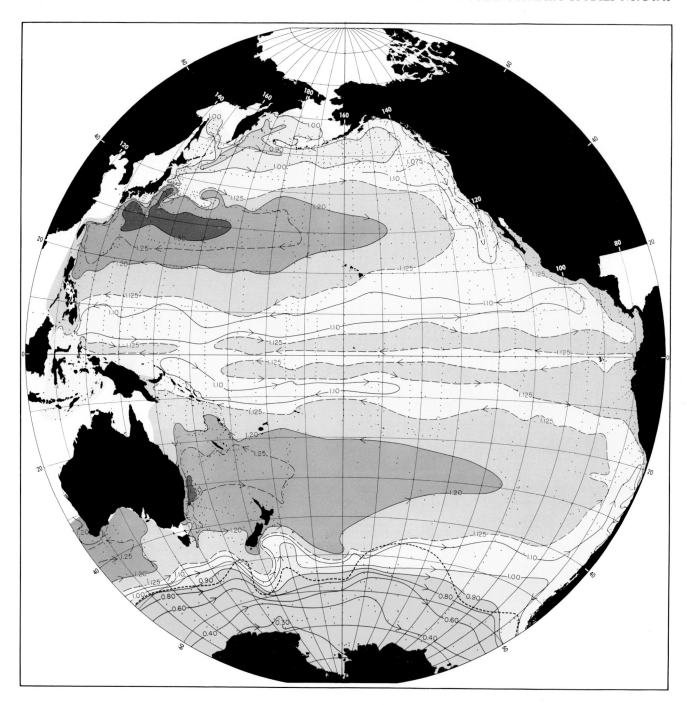

Figure 18.   Acceleration potential (from whose gradient the geostrophic flow may be calculated) on the surface where $\delta_T = 125$ cl/ton, in dynamic meters, with respect to the 1,000-db surface. The gradients vary widely and the contour intervals are not regular. The dash line near 50°S represents the intersection of the 125-cl/ton surface with the sea surface in southern summer (copied from the previous map). South of the dash line the quantity mapped is the geopotential anomaly at the sea surface with respect to the 1,000-db surface.

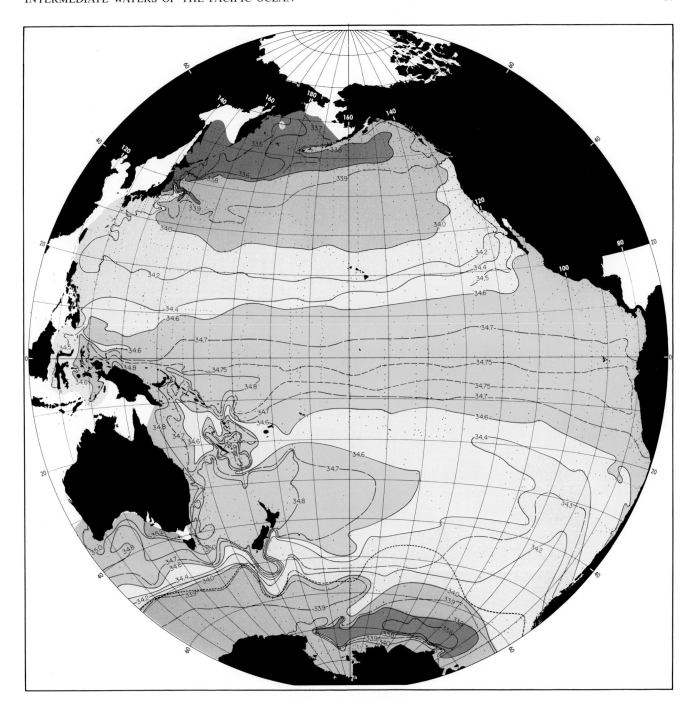

Figure 19.   Salinity (per mil) on the surface where $\delta_T = 125$ cl/ton. Corresponding values of temperature are given in table 5. South of the dash line near 50°S (the intersection of the 125-cl/ton surface with the sea surface), the quantity mapped is the salinity at the sea surface.

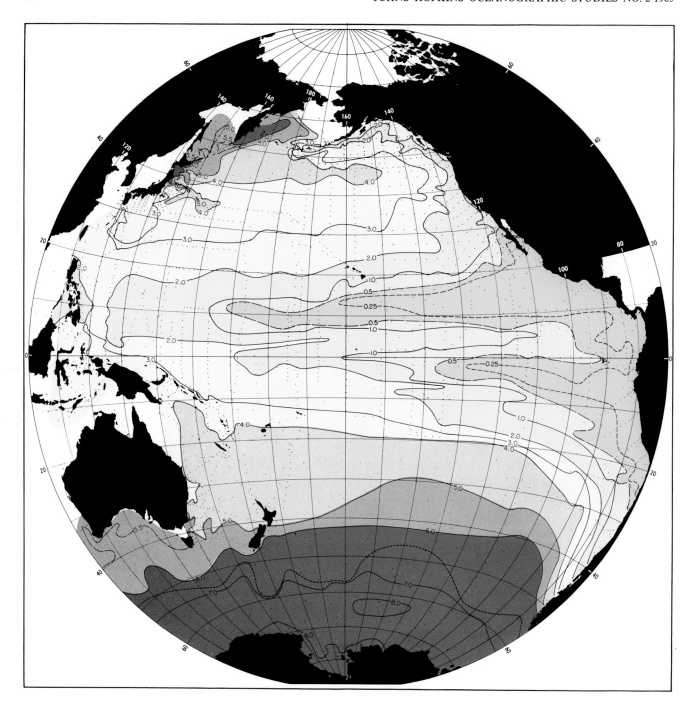

Figure 20.   Oxygen concentration (in milliliters per liter) on the surface where $\delta_T = 125$ cl/ton. South of the dash line near 50°S (the intersection of the 125-cl/ton surface with the sea surface), the quantity mapped is the oxygen concentration at the sea surface (actually 1 to 4 m).

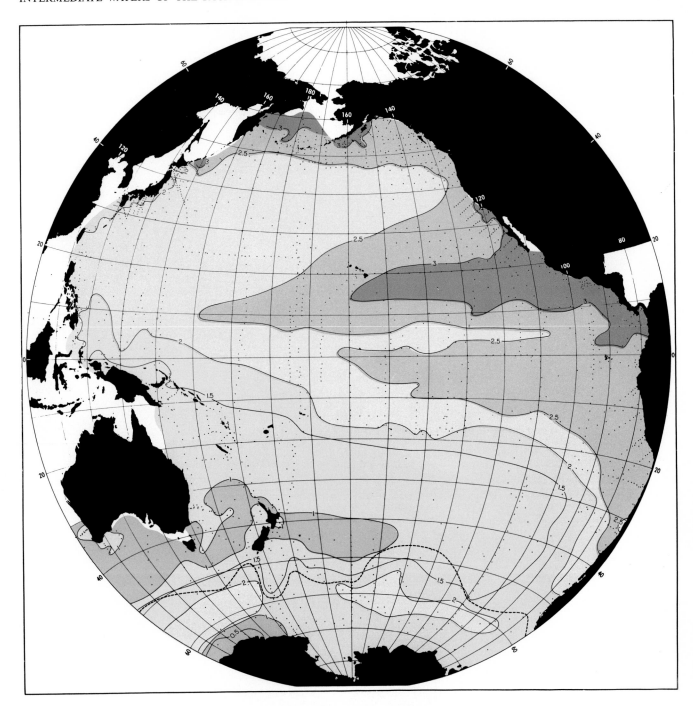

Figure 21.   Inorganic phosphate-phosphorus (in microgram-atoms per liter) on the surface where $\delta_T = 125$ cl/ton. South of the dash line near 50°S (the intersection of the 125-cl/ton surface with the sea surface), the quantity mapped is the phosphate concentration at the sea surface.

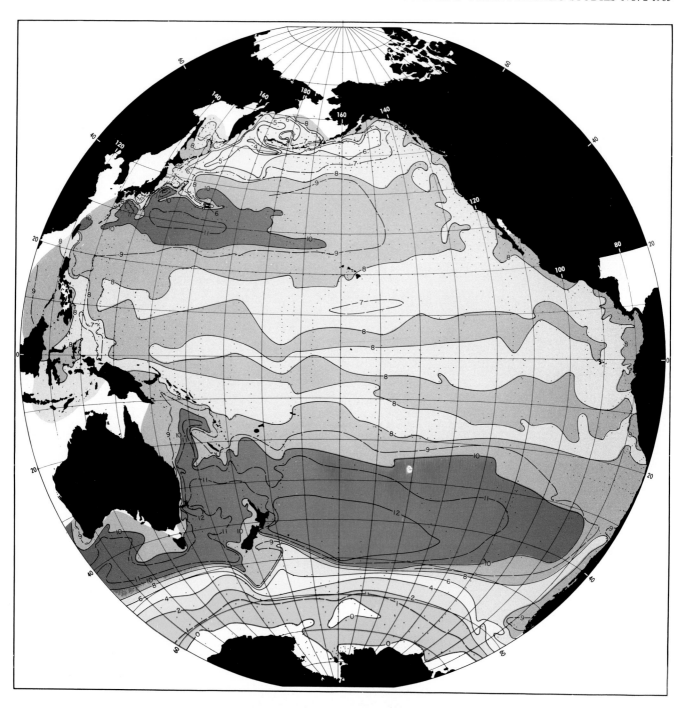

Figure 22.   Depth (in hectometers) of the surface where $\delta_T = 80$ cl/ton. The manner of preparation of the maps at 80 cl/ton is described briefly in sec. 3-2. The data sources for all the maps are given in table 3.

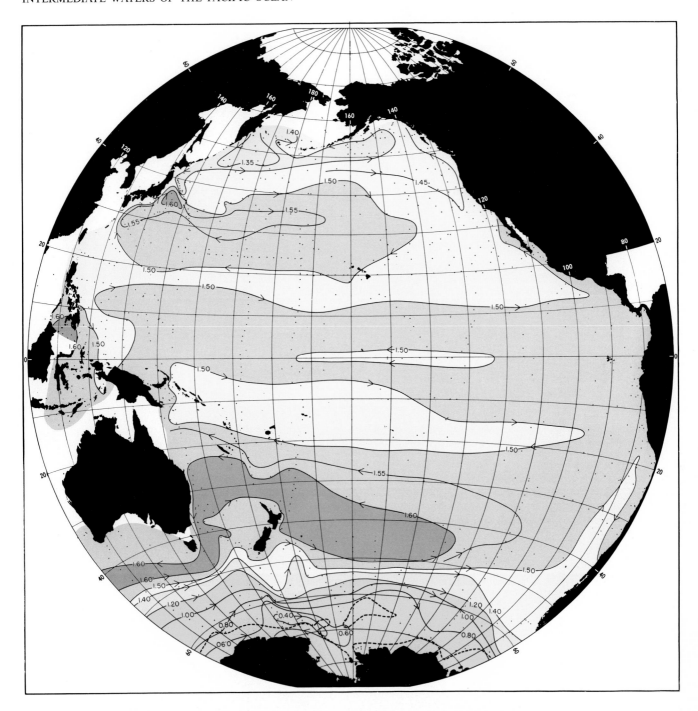

Figure 23.   Acceleration potential (from whose gradient the geostrophic flow may be calculated) on the surface where $\delta_T = 80$ cl/ton. The gradients vary widely, and the contour intervals are not regular. The dash line near Antarctica represents the intersection of the
80-cl/ton surface with the sea surface in southern summer (copied from the previous map).
South of the dash line the quantity mapped is the geopotential anomaly at the sea surface with
respect to the 2,000-db surface.

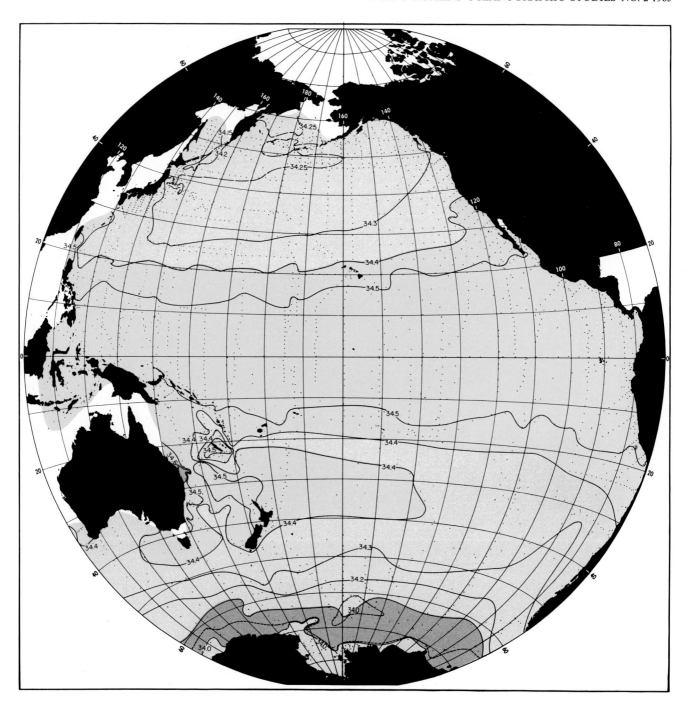

Figure 24.   Salinity (per mil) on the surface where $\delta_T = 80$ cl/ton. Corresponding values of temperature are given in table 5. South of the dash line (the intersection of the 80-cl/ton surface with the sea surface), the quantity mapped is the salinity at the sea surface.

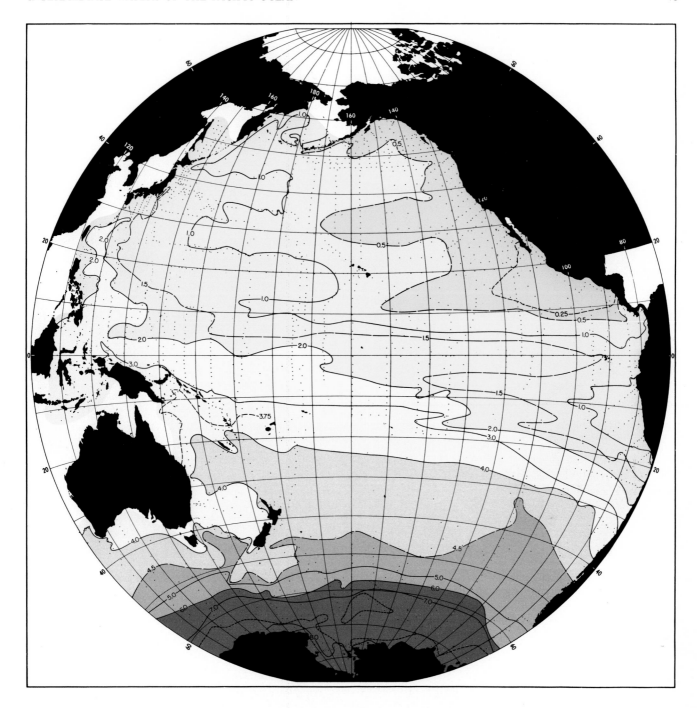

Figure 25.   Oxygen concentration (in milliliters per liter) on the surface where $\delta_T = 80$ cl/ton. South of the dash line (the intersection of the 80-cl/ton surface with the sea surface), the quantity mapped is the oxygen concentration at the sea surface (actually 1 to 4 m).

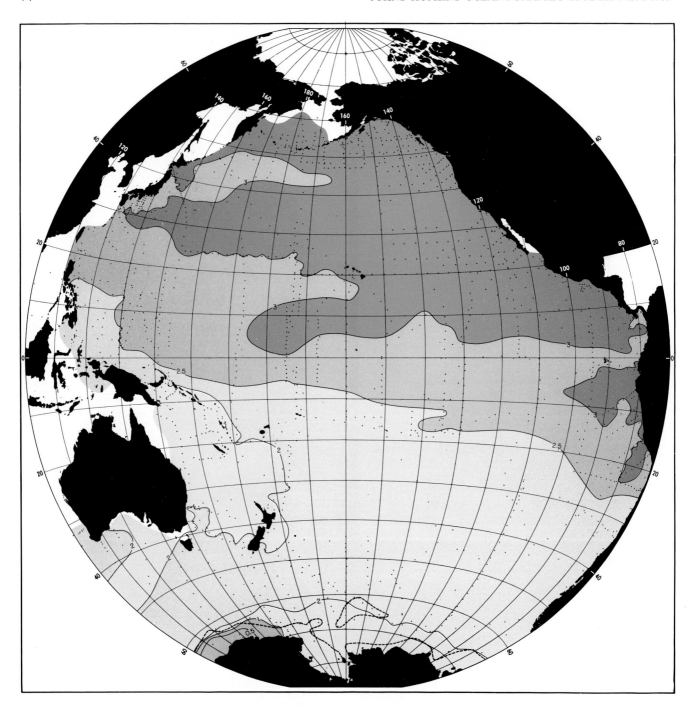

Figure 26.   Inorganic phosphate-phosphorus (in microgram-atoms per liter) on the surface
where $\delta_T = 80$ cl/ton. South of the dash line (the intersection of the 80-cl/ton surface with
the sea surface), the quantity mapped is the phosphate concentration at the sea surface.

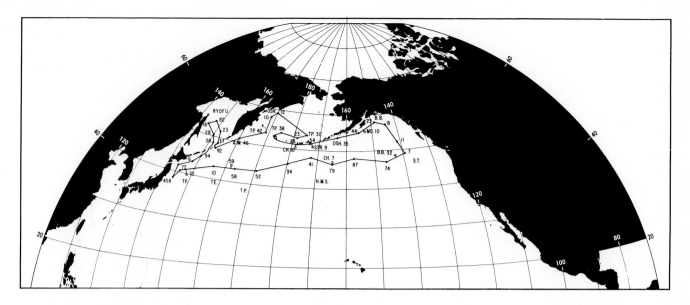

Figure 27.   Positions of the stations used in the vertical section around the subarctic gyre. Open circles indicate stations used below 1,000 m only. The manner of preparation of the sections is described briefly in sec. 3-2. Data sources are given in table 4.

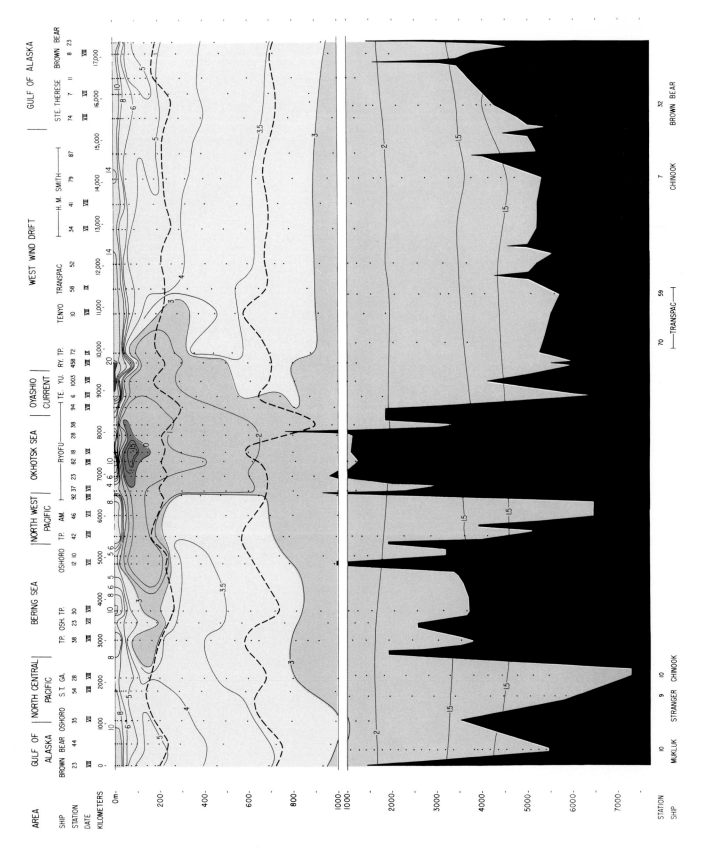

Figure 28.   Temperature (Celsius) around the subarctic gyre. Dash lines on this and the other vertical sections indicate the depths of the 125-cl/ton and 80-cl/ton isopleths.

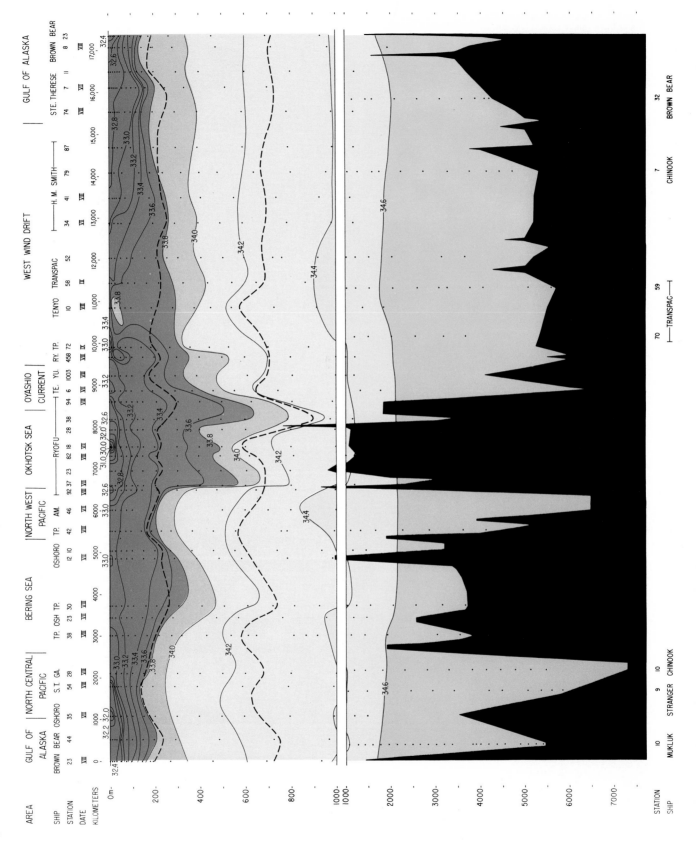

Figure 29. Salinity (per mil) around the subarctic gyre.

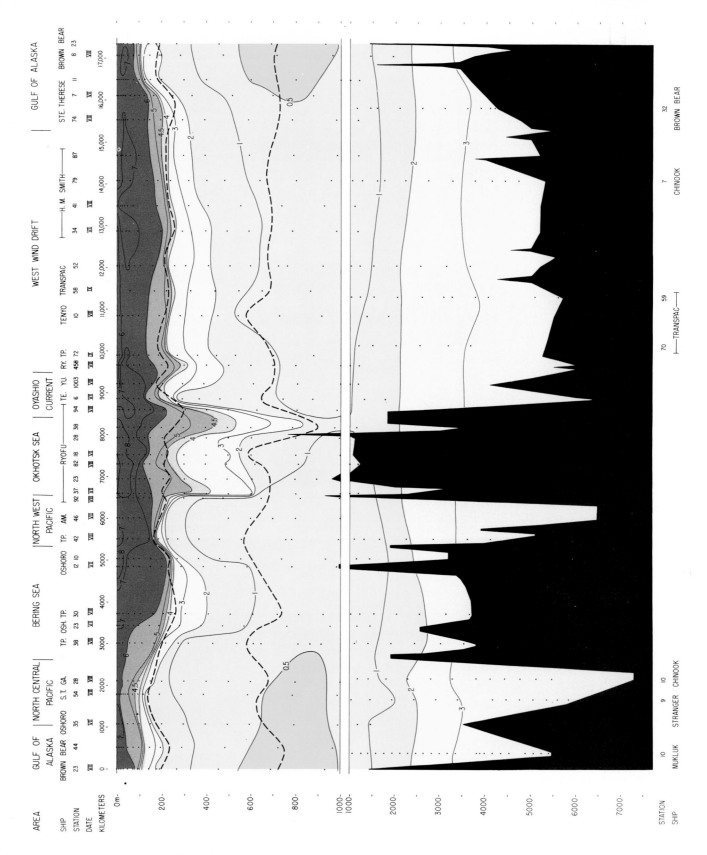

Figure 30.   Oxygen (in milliliters per liter) around the subarctic gyre.

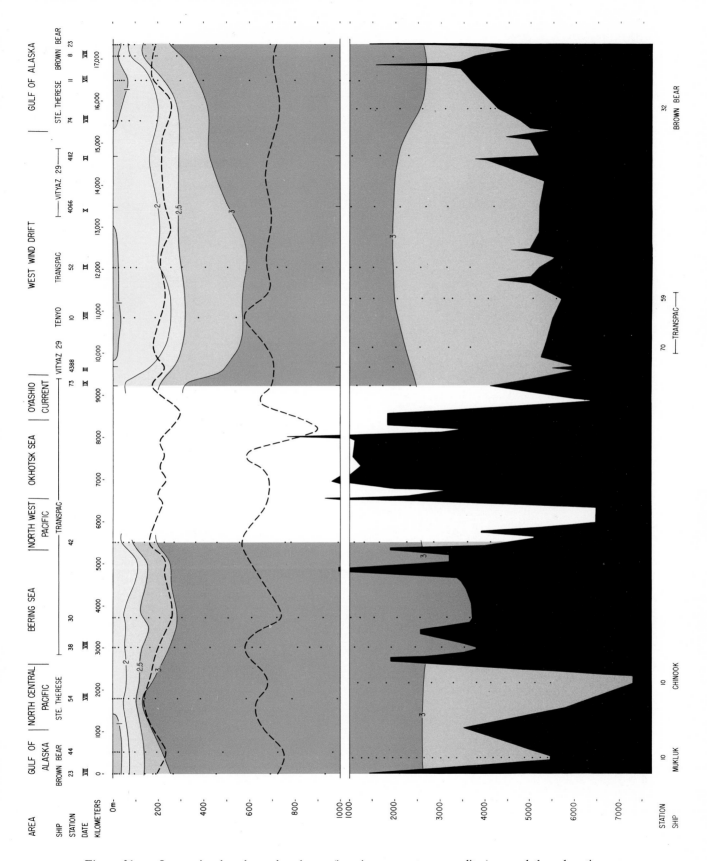

Figure 31.   Inorganic phosphate-phosphorus (in microgram-atoms per liter) around the subarctic gyre.

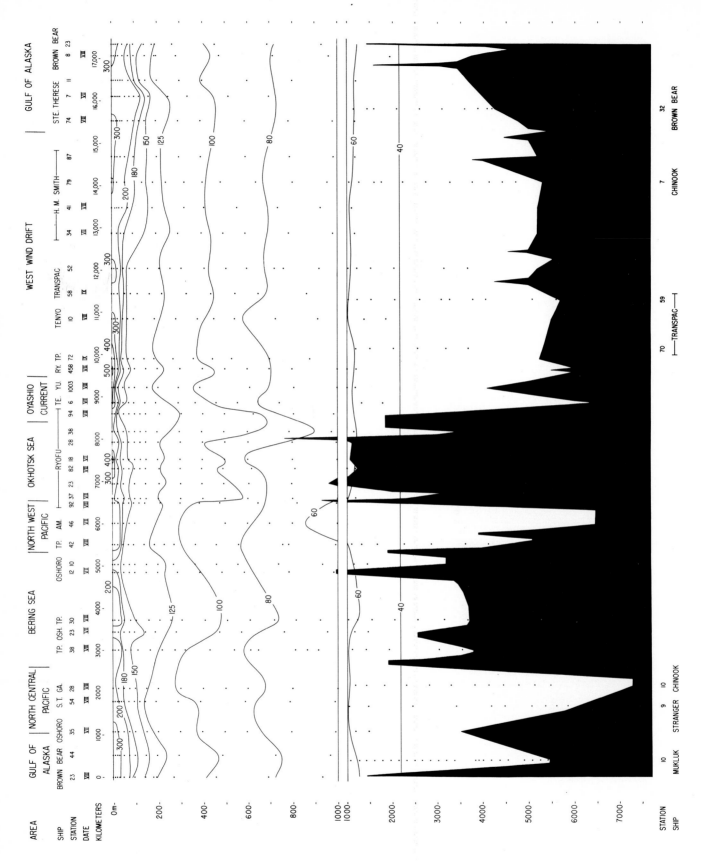

Figure 32.   Thermosteric anomaly (in centiliters per ton) around the subarctic gyre.

References

Academy of Sciences of the U.S.S.R., 1958: Hydrological, hydrochemical, geological and biological studies, research ship "Ob" 1955–1956. *IGY Reports of the complex Antarctic expedition of the Academy of Sciences of the U.S.S.R.* Leningrad, Hydro-Meteorological Publishing House, 214 p.

Agricultural Technology Association, 1954: Northern Area oceanographic data 1887–1953. *Report of the investigation of long range cold weather forecasting*, no. 1, Tokyo, 556 p.

Akagawa, Masami, 1958: On the relation between oceanographical conditions and sea ice in the Okhotsk Sea. *Bull. Hakodate Mar. met. Obs.*, **5**, 91–104.

Austin, T. S., 1957: Summary, oceanographic and fishery data, Marquesas Islands area, August–September, 1956 (EQUAPAC). *Spec. sci. Rep. U. S. Fish Wildl. Serv.*, Fish. no. **217**, 186 p.

Barnes, C. A., & T. G. Thompson, 1938: Physical and chemical investigations in Bering Sea and portions of the North Pacific Ocean. *Univ. Wash. Publ. Oceanogr.*, **3**, 35–79 & appendix.

Belknap, G. E., 1874: *Deep-sea soundings in the North Pacific Ocean, obtained in the U.S.S. Tuscarora.* U. S. Navy Hydrographic office, **54**, 51 p. & 19 plates & 9 profiles & 1 chart.

Brandhorst, Wilhelm, 1958: Thermocline topography, zooplankton standing crop, and mechanisms of fertilization in the eastern tropical Pacific. *J. Cons.*, **24**, 16–31.

Bruneau, Leif, N. G. Jerlov, & F. F. Koczy, 1953: Physical and chemical methods, Appendix Table 2. *Reports of the Swedish deep-sea expedition 1947–1948*, **3**, Physics and chemistry, no. 4, XLII–LV.

Buchanan, J. Y., 1876: *The Challenger expedition.* Papers by Sir Wyville Thompson, Mr. Murray, Mr. Mosely, Mr. Buchanan, and the late Dr. von Willemoes-suhm, Ch. VI, 593–622. London, 636 p.

———, 1877: On the distribution of salt in the ocean, as indicated by the specific gravity of its waters. *J. R. geogr. Soc.*, **47**, 72–86 & 3 folded plates.

———, 1884: Report on the specific gravity of samples of ocean water, observed on board H.M.S. *Challenger* during the years 1873–1876. In *Report on the scientific results of the voyage of H.M.S. Challenger during the years 1873–1876*, **1**, Physics and chemistry, part 2. London, H.M. Stationery Office, 1–46 & 11 diagrams.

Carlsberg Foundation, 1937: Hydrographical observations made during the DANA expedition, 1928–1930. *Dana Rep.*, **2**, 46 p.

Central Meteorological Observatory of Japan, Tokyo, 1952: *The results of marine meteorological and oceanographical observations*, no. 7, January–June 1950, 220 p.

Cochrane, J. D., 1958: The frequency distribution of water characteristics in the Pacific Ocean. *Deep-Sea Res.*, **5**, 111–27.

Commonwealth Scientific & Industrial Research Organization, Australia, Division of Fisheries, 1956: Onshore hydrological investigations in eastern and south-western Australia, 1954. *Oceanogr. Sta. List*, **24**, 54–109.

——— 1957: Onshore and oceanic hydrological investigations in eastern and south-western Australia, 1955. *Oceanogr. Sta. List*, **27**, 66–145.

———, 1962: Oceanographical observations in the Pacific Ocean in 1960, H.M.A.S. "Gascoyne," Cruises G 1/60 and G 2/60. *Oceanogr. Cruise Rep.* no. 5, 255 p. & 49 figs.

Cromwell, Townsend, 1953: Circulation in a meridional plane in the central equatorial Pacific. *J. mar. Res.*, **12**, 196–213.

———, 1954: Mid-Pacific oceanography II, transequatorial waters June–August 1950, January–March 1951. *Spec. sci. Rep. U. S. Fish Wildl. Serv.*, Fish. no. **131**, 13 p. & 32 figs. & 157 tables.

———, 1958: Thermocline topography, horizontal currents and "ridging" in the eastern tropical Pacific. *Inter-Amer. trop. Tuna Comm. Bull.*, **3**, 133–64.

Crooks, A. D., 1960: Oceanic investigations in eastern Australia, H.M.A. Ships *Queenborough, Quickmatch,* and *Warrego,* 1958. Commonwealth Scientific & Industrial Research Organization, Australia, Division of Fisheries, *Oceanogr. Sta. List,* **43,** 57 p.

Deacon, G. E. R., 1937: The hydrology of the Southern Ocean. *Discovery Rep.,* **15,** 1–124 & plates I–XLIV, Cambridge University Press.

Defant, Albert, 1936: Die Troposphäre des Atlantischen Ozeans. *Wiss. Ergebn. dtsch. atlant. Exped. "Meteor,"* **6,** I Teil, 289–411.

———, 1941a: Quantitative Untersuchungen zur Statik und Dynamik des Atlantischen Ozeans. *Wiss. Ergebn. dtsch. atlant. Exped. "Meteor,"* **6,** II Teil, 191–260, Beilagen XIX–XXIX.

———, 1941b: Die relative Topographie einzelner Druckflächen in Atlantischen Ozean. *Wiss. Ergebn, dtsch. atlant. Exped. "Meteor,"* **6,** II Teil, 183–90, Beilagen I–XVIII.

Dietrich, Günter, 1960: Temperatur-, Salzgehalts- und Sauerstoff-Verteilung auf den Schnitten von F.F.S. "Anton Dohrn" und V.F.S. "Gauss" im Internationalen Geophysikalischen Jahr 1957/1958. *Dtsch. hydrogr. Z.,* Series B (4°), Nr. 4, 103 p.

Discovery Committee, 1941: Discovery Investigations Station List, 1931–1933. *Discovery Rep.,* **21,** 226 p.

———, 1944: Discovery Investigations Station List, 1935–1937. *Discovery Rep.,* **24,** 196 p.

———, 1947: Discovery Investigations Station List, 1937–1939. *Discovery Rep.* **24,** 198–422.

———, 1949: Hydrographical observations made by R.R.S. *William Scoresby* 1931–1938. *Discovery Rep.,* **25,** 143–280.

———, 1957: Hydrographical observations made during the DISCOVERY investigations, 1950–1951. *Discovery Rep.,* **28,** 300–98.

Dodimead, A. J., 1958: Report on oceanographic investigations in the northeast Pacific Ocean during August 1956, February 1957, and August 1957. *Manuscr. Rep. Ser. oceanogr. limnol. Fish. Res. Bd. Can.,* **20,** 13 p.

———, 1961: Some features of the upper zone of the subarctic Pacific Ocean. *Bull. int. N. Pacif. Fish. Comm.,* **3,** 11–24.

———, F. W. Dobson, N. K. Chippindale, & H. J. Hollister, 1962: Oceanographic data record, North Pacific survey May 23 to July 5, 1962. *Manuscr. Rep. Ser. oceanogr. limnol. Fish. Res. Bd. Can.,* **138,** 384 p.

———, Felix Favorite, & Toshiyuki Hirano, 1963: Review of oceanography of the subarctic Pacific region. *Bull. int. N. Pacif. Fish. Comm.,* **13,** 195 p.

Equapac Committee (in press): *Oceanic observations of the Pacific: 1956, the EQUAPAC DATA.*

Fleming, J. A., H. U. Sverdrup, C. C. Ennis, S. L. Seaton, & W. C. Hendrix, 1945: Observations and results in physical oceanography, graphical and tabular summaries, *Carnegie Instn. Wash. Publ.,* **545,** Scientific results of cruise VII of the *Carnegie* during 1928–1929 under command of Captain J. P. Ault, Oceanography, I-B, 315 p.

Fleming, R. H., and staff, 1959: Cruise 199, M.V. BROWN BEAR, July to August 1958 for the International Geophysical Year of 1957–58. *Univ. Wash., Dep. Oceanogr., Spec. Rep.* no. 30, 58–32, 280 p.

Fuglister, F. C., 1960: Atlantic Ocean atlas of temperature and salinity profiles and data from the International Geophysical Year of 1957–1958. *Woods Hole Oceanographic Institution Atlas Series,* **1,** 209 p.

Hirano, Toshiyuki, 1957: The oceanographic study on the subarctic region of the northwestern Pacific Ocean (Parts 1 and II). On the water systems in the subarctic region. *Bull. Tokai Reg. Fish. Res. Lab.,* **15,** 39–70.

———, 1961: The oceanographic study on the subarctic region of the northwestern Pacific Ocean (Part IV). On the circulation of the subarctic water. *Bull. Tokai Reg. Fish. Res. Lab.,* **29,** 11–38.

Hokkaido University, Faculty of Fisheries, 1957: *Data Rec. oceanogr. Obsns. explor. Fish.,* **1,** 247 p.

———, 1958: *Data Rec. oceanogr. Obsns. explor. Fish.,* **2,** 199 p.

———, 1959: *Data Rec. oceanogr. Obsns. explor. Fish.,* **3,** 296 p.

———, 1960: *Data Rec. oceanogr. Obsns. explor. Fish.,* **4,** 221 p.

———, 1961: *Data Rec. oceanogr. Obsns. explor. Fish.,* **5,** 392 p.

———, 1962: *Data Rec. oceanogr. Obsns. explor. Fish.,* **6,** 283 p.

———, 1963: *Data Rec. oceanogr. Obsns. explor. Fish.,* **7,** 262 p.

Holmes, R. W., & Maurice Blackburn, 1960: Physical, chemical, and biological observations in the eastern tropical Pacific Ocean, SCOT expedition, April–June 1958. *Spec. sci. Rep. U. S. Fish Wildl. Serv.,* Fish. no. **345,** 106 p.

Humboldt, Alexandre de, 1831: *Fragmens de géologie et de climatologie Asiatiques,* **2,** 310–640, Paris.

Ichiye, Takashi, 1954: On the distributions of oxygen and their seasonal variations in the adjacent seas of Japan, parts I–III. *Oceanogr. Mag.,* **6,** 41–131.

———, 1955: On the possible origins of the intermediate water in the Kuroshio. *Rec. oceanogr. Wks. Jap.,* New Series, **2,** 82–89.

———, 1956: On the distributions of oxygen and their seasonal variations in the adjacent seas of Japan, part IV. *Oceanogr. Mag.,* **8,** 1–27.

————, 1962: On formation of the intermediate water in the northern Pacific Ocean. *Geofis. pur. appl.,* **51**, Milano.

Iida, Hayato, 1962: On the water masses in the coastal region of the south-western Okhotsk Sea. *J. oceanogr. Soc. Japan,* 20th Anniversary Vol., 272–78.

Institut Français d'Océanie: ORSOM III: Croisière EQUAPAC, Sept.–Oct. 1956, unpub. rep.

Institute of Oceanology, Academy of Sciences of the U.S.S.R., 1961: Hydrology, Hydrochemistry. *Data of oceanological investigations, R.V. "Vityaz," Pacific Ocean, October 1958– March 1959,* **2**, 214 p.

IGY World Data Center A, Oceanography, 1961: Hydrological observations in the southern oceans. *IGY Oceanography Report* no. 2, Department of Oceanography and Meteorology, Agricultural and Mechanical College of Texas, 386 p.

Jacobs, W. C., 1951: The energy exchange between sea and atmosphere and some of its consequences. *Bull. Scripps Instn. Oceanogr. Univ. Calif.,* **6**, 27–122.

Japan Meteorological Agency, 1960a: *The results of marine meteorological and oceanographical observations,* no. 25, January–June 1959, 258 p.

————, 1960b: *The results of marine meteorological and oceanographical observations,* no. 26, July–December 1959, 256 p.

Kagoshima University, Faculty of Fisheries, 1957: *Oceanographical observation made during the International Cooperative Expedition EQUAPAC in July–August, 1956, by M.S. Kagoshima-maru and by M.S. Keiten-maru,* 68 p.

King, J. E., T. S. Austin, & M. S. Doty, 1957: Preliminary report on Expedition EASTROPIC. *Spec. sci. Rep. U. S. Fish Wildl. Serv.,* Fish no. **201**, 155 p.

Kirwan, A. D., 1963: *Circulation of Antarctic Intermediate Water deduced through isentropic analysis.* Agricultural and Mechanical College of Texas Research Foundation, Reference 63-34F, 34 p. & 47 figs.

Kitamura, Hiroyuki, 1958: On the distribution of phosphate in the western North Pacific. *Mem. Imp. Mar. Obs. Kobe,* **12**, 1–6.

Knauss, J. A., 1960: Measurements of the Cromwell Current. *Deep-Sea Res.,* **6**, 265–86.

Koenuma, Kwan'ichi, 1936: On the hydrography of the southwestern part of the North Pacific and the Kurosio. Part I. General oceanographic features of the region. *Mem. Imp. Mar. Obs. Kobe,* **6**, 279–332.

————, 1937: On the hydrography of the southwestern part of the North Pacific and the Kurosio. Part II. Characteristic water masses which are related to this region, and their mixtures, especially the water of the Kurosio. *Mem. Imp. Mar. Obs. Kobe,* **6**, 349–414.

————, 1939: On the hydrography of the southwestern part of the North Pacific and the Kurosio. Part III. Oceanographical investigations of the Kurosio area and its outer regions; development of ocean currents in the North Pacific. *Mem. Imp. Mar. Obs. Kobe,* **7**, 41–114.

Koshliakov, M. N., 1961: On water dynamics of the northwestern part of the Pacific Ocean (in Russian). *Trud. Inst. Okeanol. Akad. Nauk SSSR,* **38**, 31–55.

Koto, Hideto, & Takeji Fujii, 1958: Structure of the waters in the Bering Sea and the Aleutian region. *Bull. Fac. Fish. Hokkaido Univ.,* **9**, 149–70.

Kuksa, V. I., 1962: On the formation and distribution of the layer of low salinity in the northern part of the Pacific Ocean (in Russian). *Okeanologiia,* **2**, 769–82.

————, 1963: Basic regularities in formation and distribution of intermediate waters in the northern part of the Pacific Ocean (in Russian). *Okeanologiia,* **3**, 30–43.

LaFond, E. C., 1951: Processing oceanographic data. *H.O. Pub.* no. **614**, U. S. Navy Hydrographic Office, 114 p.

Lindenkohl, Adolphus, 1897: Das spezifische Gewicht des Meerwassers im Nordost-Pazifischen Ozean im Zusammenhange mit Temperatur- und Strömungszustanden. *Petermanns geogr. Mitt.,* **12**, 273–79.

Makaroff, Stepan, 1894: *Le "Vitiaz" et l'Océan Pacifique.* **1**, 337 p.; **2**, 511 p., plates 1–31, St. Petersburg.

Maritime Safety Agency, Tokyo, Japan, 1950: The results of oceanographic observation in the northwestern Pacific Ocean, no. 3, 1931–1935. Pub. No. 981, *Hydrogr. Bull., Tokyo,* Special Number, 139 p.

————, 1951: The results of oceanographic observation in the northwestern Pacific Ocean, no. 4, 1935–1938. Pub. No. 981, *Hydrogr. Bull., Tokyo,* Special Number, No. 8, 143 p.

————, 1952: The results of oceanographic observation in the northwestern Pacific Ocean, no. 5, 1938–1941. Pub. No. 981, *Hydrogr. Bull., Tokyo,* Special Number, No. 9, 194 p.

Maritime Safety Board, Tokyo, Japan, 1962: The results of oceanographic observation in the northwestern Pacific Ocean, no. 8, 1931–1940. Pub. No. 981, *Hydrogr. Bull., Tokyo,* No. 69, 245 p.

Masuzawa, Jotaro, 1950: On the intermediate water in the southern Sea of Japan. *Oceanogr. Mag.,* **2**, 137–44.

McGary, J. W., & E. D. Stroup, 1956: Mid-Pacific oceanography, part VIII, middle latitude waters, January–March 1954. *Spec. sci. Rep. U. S. Fish Wildl. Serv.,* Fish no. **180**, 173 p.

Mishima, Seikichi, & Satoshi Nishizawa, 1955: Report on hydrographic investigations in Aleutian waters and the southern Bering Sea in the early summers of 1953 and 1954. *Bull. Fac. Fish. Hokkaido Univ.,* **6**, 85–124.

Montgomery, R. B., 1937: A suggested method for representing gradient flow in isentropic surfaces. *Bull. Amer. meteor. Soc.,* **18,** 210–12.

———, 1938: Circulation in the upper layers of southern North Atlantic deduced with use of isentropic analysis. *Pap. phys. Oceanogr. & Met.,* **6,** published by Massachusetts Institute of Technology and Woods Hole Oceanographic Institution, 55 p.

———, 1939: Ein Versuch, den vertikalen und seitlichen Austausch in der Tiefe der Sprungschicht im äquatorialen Atlantischen Ozean zu bestimmen. *Ann. Hydrogr. mar. Meteor., Berl.,* **67,** 242–46.

———, & M. J. Pollak, 1942: Sigma-t surfaces in the Atlantic Ocean. *J. mar. Res.,* **5,** 20–27.

———, 1954: Analysis of a *Hugh M. Smith* oceanographic section from Honolulu southward across the equator. *J. mar. Res.,* **13,** 67–75.

———, & W. S. Wooster, 1954: Thermosteric anomaly and the analysis of serial oceanographic data. *Deep-Sea Res.,* **2,** 63–70.

———, & E. D. Stroup, 1962: Equatorial waters and currents at 150°W in July–August 1952. *Johns Hopkins Oceanographic Studies,* **1,** 68 p.

Norpac Committee, 1960a: *Oceanic observations of the Pacific: 1955, the NORPAC ATLAS.* Berkeley and Tokyo, University of California Press and University of Tokyo Press, 123 plates.

Norpac Committee, 1960b: *Oceanic observations of the Pacific: 1955, the NORPAC DATA.* Berkeley and Tokyo, University of California Press and University of Tokyo Press, 532 p.

Officers of the Challenger Expedition, 1884: Report on the deep-sea temperature observations of ocean water. *Report on the scientific results of the voyage of H.M.S. Challenger during the years 1873–76,* **1,** Physics and chemistry, part 3, plates I–CCLVIII & tables I–VII.

Parr, A. E., 1938: Isopycnic analysis of current flow by means of identifying properties. *J. mar. Res.,* **1,** 133–54.

Postma, Henry, 1959: Tables. Oxygen, hydrogen ion, alkalinity and phosphate. *Snellius-Exped.,* **4,** 35 p.

Prestwich, Joseph, 1875: Tables of temperatures of the sea at different depths beneath the surface, reduced and collated from the various observations made between the years 1749 and 1868, discussed. *Phil. Trans.,* **165,** Part II, 587–674 & plates 65–68.

Reichar, A. C., 1911: Temperatur- und Salzgehaltsbestimmungen im sudwestlichen Stillen Ozean, 1910. *Ann. Hydrogr. mar. Meteor., Berl.,* **XXXIX,** Heft X, 521–27.

Reid, J. L., Jr., G. I. Roden, & J. G. Wyllie, 1958: Studies of the California Current system. *Progr. Rep. Calif. Coop. ocean. Fish. Invest.,* 1 July 1956 to 1 Jan. 1958, 28–56.

———, 1959: Evidence of a South Equatorial Countercurrent in the Pacific Ocean. *Nature, London,* **184,** 209–10.

———, 1961a: On the formation and movement of the North Pacific Intermediate Water. *Pacif. Sci. Cong.,* 10th, Honolulu, Abstr. Symp. Pap., p. 345.

———, 1961b: On the geostrophic flow at the surface of the Pacific Ocean with respect to the 1,000-decibar surface. *Tellus,* **13,** 489–502.

———, 1961c: On the temperature, salinity and density differences between the Atlantic and Pacific Oceans in the upper kilometre. *Deep-Sea Res.,* **7,** 265–75.

———, 1962a: Distribution of dissolved oxygen in the summer thermocline. *J. mar. Res.,* **20,** 138–48.

———, 1962b: On the circulation, phosphate-phosphorus content and zooplankton volumes in the upper part of the Pacific Ocean. *Limnol. & Oceanogr.,* **7,** 287–306.

Riley, G. A., 1951: Oxygen, phosphate, and nitrate in the Atlantic Ocean. *Bull. Bingham oceanogr. Coll.,* **13** (1), 126 p.

Rochford, D. J., 1960a: The intermediate depth waters of the Tasman and Coral Seas. I. The 27.20 σt surface. *Aust. J. Mar. Freshw. Res.,* **11,** 127–47.

———, 1960b: The intermediate depth waters of the Tasman and Coral seas. II. The 26.80 σt surface. *Aust. J. Mar. Freshw. Res.,* **11,** 148–65.

Roden, G. I., & G. W. Groves, 1959: Recent oceanographic investigations in the Gulf of California. *J. mar. Res.,* **18,** 10–35.

Rossby, C. G., 1936: Dynamics of steady ocean currents in the light of experimental fluid mechanics. *Pap. phys. Oceanogr. & Met.,* **5,** published by Massachusetts Institute of Technology and Woods Hole Oceanographic Institution, 43 p.

Rotschi, Henri, 1958a: ORSOM III, croisière "Astrolabe," océanographie physique. *Rapp. sci. Inst. franç. Océan., Nouméa,* **8,** 79 p.

———, 1958b: Resultats des observations scientifiques du "Tiare," croisière "Bounty," 20–29 Juin 1958. *Rapp. sci. Inst. franç. Océan., Nouméa,* **7,** 20 p.

———, 1958c: ORSOM III, océanographie physique, rapport technique de la croisière 56-5. *Rapp. sci. Inst. franç. Océan., Nouméa,* **5,** 34 p.

———, 1960: ORSOM III, resultats de la croisière "Dillon," Ière partie, océanographie physique. *Rapp. sci. Inst. franç. Océan., Nouméa,* **18,** 58 p.

Rumford, Benjamin, Count of, 1798: Of the propagation of heat in fluids. *Essays, political, economical, and philosophical.* **II,** Essay VII, Part I, 199–310, London.

Saito, Yukimaso, 1952: On the Oyashio Current (Part I), a synthetic research on the oceanographical state of the North Pacific Ocean. *J. Inst. Polyt., Osaka* (B), **3,** 79–140.

Schott, Gerhard, 1935: *Geographie des Indischen und Stillen Ozeans.* Hamburg, 413 p. & 37 charts & 114 figs.

Scripps Institution of Oceanography, University of California, 1957: *Data collected by Scripps Institution vessels on EQUAPAC expedition, August 1956.* SIO Ref. 57-24, 111 p.

———, 1960: *Oceanic observations of the Pacific: 1950.* Berkeley and Los Angeles, University of California Press, 536 p.

———, 1961a: *Oceanic observations of the Pacific: Pre-1949.* Berkeley and Los Angeles, University of California Press, 349 p.

———, 1961b: *Preliminary Rep. Part I. Phys. & chem. data. STEP-I expedition 15 Sept.–14 Dec. 1960.* SIO Ref. 61-9, 48 p.

———, 1962: *Oceanic observations of the Pacific: 1955.* Berkeley and Los Angeles, University of California Press, 477 p.

———, 1963a: *Oceanic observations of the Pacific: 1951.* Berkeley and Los Angeles, University of California Press, 598 p.

———, 1963b: *Oceanic observations of the Pacific: 1956.* Berkeley and Los Angeles, University of California Press, 458 p.

———, 1965a: *Oceanic observations of the Pacific: 1953.* Berkeley and Los Angeles, University of California Press, 576 p.

———, 1965b: *Oceanic observations of the Pacific: 1957.* Berkeley and Los Angeles, University of California Press, 707 p.

———, 1965c: *Oceanic observations of the Pacific: 1958.* Berkeley and Los Angeles, University of California Press, 804 p.

———, 1965d: *Oceanic observations of the Pacific: 1952.* Berkeley and Los Angeles, University of California Press, 617 p.

———, 1965e: *Oceanic observations of the Pacific: 1959.* Berkeley and Los Angeles, University of California Press, 901 p.

Service Hydrographique de la Marine, 1960: Stations hydrologiques de patrouilleur "Lotus" de la Marine Nationale (1957–58). *Cah. Océanogr.,* **12,** 590–610.

Suda, Kwanji: EQUAPAC, oceanographic and meteorological data. Japanese Hydrographic Office, unpub. rep.

Sverdrup, H. U., 1931: The origin of the Deep-Water of the Pacific Ocean as indicated by the oceanographic work of the Carnegie. *Beitr. Geophys.,* **29,** 95–105.

———, 1939: Lateral mixing in the deep water of the South Atlantic Ocean, *J. mar. Res.,* **2,** 195–207.

———, & R. H. Fleming, 1941: The waters off the coast of Southern California, March to July, 1937. *Bull. Scripps Instn. Oceanogr. Univ. Calif.,* **4,** 267–378.

———, M. W. Johnson, & R. H. Fleming, 1942: *The oceans, their physics, chemistry, and general biology.* New York, Prentice-Hall, 1087 p.

Taft, B. A., 1963: Distribution of salinity and dissolved oxygen on surfaces of uniform potential specific-volume in the South Atlantic, South Pacific, and Indian Oceans. *J. mar. Res..,* **21,** 129–46.

Thompson, T. G., 1932: Physical properties of sea water. Physics of the earth, **5,** Oceanography, p. 63–94. *National Research Coun. Bull.,* no. 185, Washington, D.C.

Tully, J. P., 1957: Some characteristics of sea water structure. *Proc. Pacif. Sci. Cong.,* 8th, Philippines, **3,** Oceanography, p. 643–61.

———, & F. G. Barber, 1960: An estuarine analogy in the subarctic Pacific Ocean. *J. Fish. Res. Bd. Can.,* **17,** 91–112.

Uda, Michitaka, 1935: On the distribution, formation and movement of the dicho-thermal water to the northeast of Japan (in Japanese). *Umi to Sora,* **15,** 445–52.

———, 1938: Researches on "Siome" or current rip in the seas and oceans. *Geophys. Mag., Tokyo,* **11,** 307–72.

———, 1963: Oceanography of the Subarctic Pacific Ocean. *J. Fish. Res. Bd. Can.,* **20,** 119–79.

U. S. Navy Hydrographic Office, 1958: Oceanographic survey results Bering Sea area winter and spring 1955. *Tech. Rep. U. S. Navy hydrogr. Off.,* **56,** 95 p.

———, 1962: Operation DEEP FREEZE 61, 1960–1961. *Marine geophys. investigations,* TR-105, 217 p.

Van Riel, P. M., H. C. Hamaker, & L. Van Eyck, 1950: Oceanographic results, part 6, tables serial and bottom observations, temperature, salinity and density. *Snellius-Exped.,* **2,** 41 p.

Wilson, R. C., & M. O. Rinkel, 1957: Marquesas area oceanographic and fishery data, Jan.–March 1957. *Spec. sci. Rep. U. S. Fish Wildl. Serv.,* Fish. no. **238,** 136 p.

Wooster, W. S., & Townsend Cromwell, 1958: An oceanographic description of the eastern tropical Pacific. *Bull. Scripps Instn. Oceanogr. Univ. Calif.,* **7,** 169–282.

———, 1961: Further evidence of a Pacific south equatorial countercurrent. *Deep-Sea Res.,* **8,** 294–97.

Wüst, Georg, 1929: Schichtung und Tiefenzirkulation des Pazifischen Ozeans. *Veröff. Inst. Meeresk. Univ. Berlin,* neue Folge, A, **20,** 64 p.

———, 1935: Die Stratosphäre. *Wiss. Ergebn. dtsch. atlant. Exped. "Meteor,"* **6,** I Teil, 109–288.

———, & Albert Defant, 1936: Atlas zur Schichtung und Zirkulation des Atlantischen Ozeans. Schnitte und Karten von Temperatur, Salzgehalt und Dichte. *Wiss. Ergebn. dtsch. atlant. Exped. "Meteor,"* **6,** Atlas, CIII Beilagen.

Wyrtki, Klaus, 1961: Physical oceanography of the Southeast Asian Waters. Scientific results of marine investigations of the South China Sea and the Gulf of Thailand 1959–1961, *Naga Rep.,* **2,** 195 p.

———, 1962: The subsurface water masses in the western South Pacific Ocean. *Aust. J. Mar. Freshw. Res.,* **13,** 18–47.